Environments, Risks and Health

Much of the scientific work on environmental health research has come from the clinical and biophysical sciences. Yet contributions are being made from the social sciences with respect to economic change, distributional equities, political will, public perceptions and the social geographical challenges of the human health–environments linkages.

Offering the first comprehensive and cohesive summary of the input from social science to this field, this book focuses on how humans theorize their relationships to the environment with respect to health and how these ideas are mediated through an evaluation of risk and hazards. Most work on risk has focused primarily on environmental problems. This book extends and synthesizes these works for the field of human health, treating social, economic, cultural and political context as vital. Bringing disparate literatures from across several disciplines together with their own applied research and experience, Eyles and Baxter deal with scientific uncertainty in the everyday issues raised and question how social theories and models of the way the world works can contribute to understanding these uncertainties.

This book is essential reading for those studying and researching in the fields of health geography and environmental studies as well as environmental sociology, social and applied anthropology, environmental psychology and environmental politics.

John Eyles is a Distinguished University Professor at McMaster University, Canada. Based in Geography and Earth Sciences, he holds appointments in Clinical Epidemiology and Biostatistics, Sociology and the Centre for Health Economics and Policy analysis. His main research interests lie in environmental influences on human health and access to health care resources within a policy context. He has pursued that latter interest through a National Research Foundation South African Research Chair in Health Policy and Systems in the Centre for Health Policy, School of Public Health, University of Witwatersrand, where he is a Distinguished Research Professor.

Jamie Baxter is Professor in the Department of Geography at Western University, Canada. His research interests include: the social construction of risks from technological hazards, community responses to hazards, environment and health, noxious facility siting, and social science research methodology.

Geographies of Health

Series Editors:

Allison Williams

Associate Professor, School of Geography and Earth Sciences, McMaster University, Canada

Susan Elliott

Professor, Department of Geography and Environmental Management and School of Public Health and Health Systems, University of Waterloo, Canada

There is growing interest in the geographies of health and a continued interest in what has more traditionally been labeled medical geography. The traditional focus of 'medical geography' on areas such as disease ecology, health service provision and disease mapping (all of which continue to reflect a mainly quantitative approach to inquiry) has evolved to a focus on a broader, theoretically informed epistemology of health geographies in an expanded international reach. As a result, we now find this subdiscipline characterized by a strongly theoretically-informed research agenda, embracing a range of methods (quantitative; qualitative and the integration of the two) of inquiry concerned with questions of: risk; representation and meaning; inequality and power; culture and difference, among others. Health mapping and modeling has simultaneously been strengthened by the technical advances made in multilevel modeling, advanced spatial analytic methods and GIS, while further engaging in questions related to health inequalities, population health and environmental degradation.

This series publishes superior quality research monographs and edited collections representing contemporary applications in the field; this encompasses original research as well as advances in methods, techniques and theories. The *Geographies of Health* series will capture the interest of a broad body of scholars, within the social sciences, the health sciences and beyond.

Also in the series:

The Afterlives of the Psychiatric Asylum Recycling Concepts, Sites and Memories
Edited by Graham Moon, Robin Kearns and Alun Joseph

Geographies of Health and Development
Edited by Isaac Luginaah and Rachel Bezner Kerr

Soundscapes of Wellbeing in Popular Music
Gavin J. Andrews, Paul Kingsbury and Robin Kearns

Mobilities and Health
Anthony C. Gatrell

Environments, Risks and Health

Social perspectives

John Eyles and Jamie Baxter

LONDON AND NEW YORK

First published 2017 by Routledge

2 Park Square, Milton Park, Abingdon, Oxfordshire OX14 4RN
52 Vanderbilt Avenue, New York, NY 10017

Routledge is an imprint of the Taylor & Francis Group, an informa business

First issued in paperback 2020

British Library Cataloguing in Publication Data
A catalogue record for this book is available from the British Library

Library of Congress Cataloging-in-Publication Data
Names: Eyles, John, author. | Baxter, Jamie, author.
Title: Environments, risks and health : social perspectives / John Eyles and Jamie Baxter.
Description: Abingdon, Oxon ; New York NY : Routledge, 2017. | Includes bibliographical references.
Identifiers: LCCN 2016009982 | ISBN 9781472410191 (hardback) | ISBN 9781315580081 (e-book)
Subjects: | MESH: Environmental Health | Risk Assessment
Classification: LCC RA565 | NLM WA 30.5 | DDC 616.9/8—dc23
LC record available at http://lccn.loc.gov/2016009982

ISBN: 978-1-4724-1019-1 (hbk)
ISBN: 978-0-367-66813-6 (pbk)

Typeset in Times New Roman
by Apex CoVantage, LLC

For Emily, with love and admiration – an engaged scholar who will become a better geographer and scientist than I. (JDE)

With love for Donna, Clare and Duncan – for saving it until we are all at the dinner table. (JWB)

Contents

Figures

Tables

Abbreviations

AAAS	American Association for the Advancement of Science
ABCC	Atom Bomb Casualty Commission
ACF	advocacy coalition framework
ADR	alternative dispute resolution
AIDS	acquired immune deficiency syndrome
ALARA	as low as reasonably achievable
ATSDR	Agency for Toxic Substances and Disease Registry
BP	British Petroleum
BPA	bisphenol A
BSE	bovine spongiform encephalopathy, or 'mad cow disease'
BTEX	benzene, toluene, ethylbenzene and xylenes
CBC	Canadian Broadcasting Corporation
CD	compact disc
CDC	Centers for Disease Control
CEA	cumulative effects assessment
CERCLA	Comprehensive Environmental Response, Compensation, and Liability Act
CFC	chlorofluorocarbon
CI	confidence interval
CIS	Commonwealth of Independent States
CRED	Centre for Research on the Epidemiology of Disasters
DALY	disability-adjusted life year
DDE	dichlorodiphenyldichloroethylene
DDT	dichlorodiphenyltrichloroethane
DEHP	bis(2-ethylhexyl) phthalate
DVD	digital versatile disk
EA	environmental assessment
ECR	environmental conflict resolution
EDB	environmental disease burden
EEG	electroencephalogram
EHIA	environmental impact assessment for health
EIA	environmental impact assessment
EMF	electromagnetic fields

EPA	Environmental Protection Agency
FDA	Food and Drug Administration
FEMA	Federal Emergency Management Agency
FEV1	forced expiratory volume (after one second)
FVC	forced vital capacity
GI	gastrointestinal
GLWQA	Great Lakes Water Quality Agreement
GMO	genetically modified organism
HCB	hexachlorobenzene
HIA	health impact assessment
HIV	human immunodeficiency virus
IARC	International Agency for Research on Cancer
ICJB	International Campaign for Justice in Bhopal
IJC	International Joint Commission
IPCC	Intergovernmental Panel on Climate Change
JECFA	Joint Expert Committee on Food Additives
KANUPP	Karachi Nuclear Power Plant
LD50	lethal dose fifty (50% of animals die)
LGA	local government association
LOAEL	lowest observable adverse effects level
LOEL	lowest observed effect level
MAUP	modifiable areal unit problem
MCF-7	Michigan Cancer Foundation-7
MeHg	methylmercury
MIC	methyl isocyanate
MPM	malignant pleural mesothelioma
MTO	Moving to Opportunity
NIMBY	not in my backyard
NOAEL	no observed adverse effects level
NOEL	no observed effect level
NRC	National Research Council
NTD	neural tube defects
OCP	organochlorine pesticides
OECD	Organisation for Economic Co-operation and Development
OMA	Ontario Medical Association
OSHA	Occupational Safety and Health Administration
PBB	polybrominated biphenyl
PCB	polychlorinated biphenyl
POP	persistent organic pollutant
PPAR	peroxisome proliferator–activated receptor
PVC	polyvinyl chloride
RAND	Research and Development Corporation
RfD	reference dose
RF-EMF	radio-frequency electromagnetic fields
SARF	social amplification (and attenuation) of risk framework

SARS	severe acute respiratory syndrome
SCCA	South Camden Citizens in Action
SEAB	Secretary of Energy Advisory Board
SHS	second-hand smoke
SWB	subjective well-being
TCDD	tetrachlorodibenzo-p-dioxin
TCE	trichloroethylene
TRI	toxic release inventory
TSP	total suspended particulates
UCC	United Church of Christ
UF, UFA	uncertainty factor
UFDB	uncertainty factor due to a database
UFFI	urea formaldehyde foam insulation
UFH	uncertainty factor due to human intraspecies variability
UFL	uncertainty factor due to difference between LOAEL and NOAEL when LOAEL used
UFS	uncertainty factor due to difference between subchronic and chronic exposure when subchronic used
UNCED	United Nations Conference on Environment and Development
UNECE	United Nations Economic Commission for Europe
UNEP	United Nations Environmental Program
UNFCCC	United Nations Framework Convention on Climate Change
USGA	United States Government Accounting Office
UVB	ultraviolet B (radiation)
vCJD	variant Creutzfeldt-Jakob Disease
VOC	volatile organic compound
WHO	World Health Organization
WTE	waste-to-energy
β-HCH	β-hexachlorocyclohexane
γ-HCH	γ-hexachlorocyclohexane

Preface

As teachers of courses at the undergraduate and graduate levels on environmental health from a social science perspective, we have felt the need for a book such as this for a while. Existing texts on environment and health are limited in terms of their coverage of the social sciences, organized instead in terms of core health sciences (toxicology and epidemiology) and the routes of exposure (air, land, water, waste). In our searches for papers to populate these courses, we are served well by epidemiology and economics but less so by sociology and political science. It may be cheating to put epidemiology in this list as it uses social variables in a specific, usually quantified form, providing insightful analyses of exposure-outcome-mediating relationships. So too does toxicology and risk assessment, but with respect to the social sciences, there are many analyses of environment from discourse analysis to ethnographic and phenomenological understandings of inequitable impact. Likewise, there are many analyses of health and variations in its distribution, measured in different ways and associated with factors which shape these distributions – age, gender, income and so on. But these health analyses often define and describe the social as an 'environment', a perspective that does not adequately account for social processes and social structures. The 'social' is not merely a variable or collection thereof. The biophysical environment is, likewise, very difficult to measure, perhaps even to conceptualize. Health is seen as individual, while environment is seen as a more abstract set of forces that affects whole populations as well as individuals. Linkage is often difficult in epidemiological and toxicological sciences. More so in the social sciences where outcomes are often reactive and initially psychosocial and study designs are usually qualitative or mixed method and cross-sectional.

To this problem, we have added risk, a multi-dimensional concept which is perceived as real by individuals and communities in specific, often common, situations in which 'something out there happens' (e.g., a fire, a leak, a spill, environmental losses and destruction of environmental assets, a proposed by-law, clusters of symptoms in residents, a siting announcement). Quantitatively and epidemiologically these circumstances may not register as risky, but they must be recognized as real and managed. We have both worked in applied settings where these things happen. For Eyles, this has been working with usually local governments, particularly public health departments and citizen groups. Most of this work has been in Ontario, Canada, and has involved tire and chemical fires, trihalomethane exposure, PCB contamination, living close to waste sites and nuclear power generating plants and potential impacts of orphan industrial sites. He has also worked overseas on similar issues – the desertification and polluting of the Aral Sea, health and social impacts of coral degradation in Indonesia and exposure to naturally occurring radioactive material disturbed by mining in south India. Baxter has likewise conducted research with communities living with waste sites – including municipal solid waste and hazardous waste that gets incinerated, landfilled or stored for transfer. He has also studied how communities sort through risk information to set municipal policy regarding urban pesticide use. More recently, he has studied difficult-to-site alternative energy facilities meant to reduce greenhouse gas emissions including energy-from-waste and wind turbine developments. All of this research involves comparative case-study designs that explore similarities and differences among communities in Alberta, Nova Scotia and Ontario.

In all these settings, the toxicological and epidemiological evidence, with the parameters of scientific uncertainty, demonstrated little or no physical health impact. But in most settings, there was significant experiential uncertainty – events, substances, and facilities were seen as risky – which resulted in impacts on psychosocial health. For what is now now a declining number of scientists these adverse outcomes were 'blamed' on individuals as being worriers – NIMBYs – who did not understand science or, worse, were being selfish by wanting socially needed facilities kept out of their backyards. It was all in the mind, but there is recognition that that is itself important. Invisible trauma impacts lifescape and the growing distrust of agencies and organizations, which explains why these anxieties matter for our overall understanding of environmental hazards in that they have a meaningful impact on health and well-being. Yet the application of rigorous methodological tests on purported environment and health relationships can often lead to uncertainty about the relationship between exposure and outcome. These local investigations are usually in real time with concerned residents and officials who are 'biased' – colouring how they see exposure which departs from a normal way of life as detrimental. But science is biased too, for example, towards a conservatism that can be at odds with precautionary policy. Furthermore, most of these local cases involve small numbers of individuals, which may be limited further by a scientific design which often uses distance from the exposure as a proxy for dose response. It is also fair to point out that most of the exposures are localized, while wider impacts such as global warming, acid rain and the distribution of pollutants to affect the Arctic or Sahel are so interconnected and produce outcomes similar to local exposure that associations are difficult to tease apart. The localized exposures usually lead to rational ways of coping with and managing the exposure which may be unheeded elsewhere or have adverse consequences in other places. There may be then a tyranny of small decisions. Kahn (1966) describes a situation in which a number of decisions, individually small in size and with a limited time perspective, cumulatively result in an outcome which is neither optimal nor desired. It is a situation in which a series of small, individually rational decisions can negatively change the context of subsequent choices, even to the point that newly valued and desired alternatives are irreversibly destroyed. Global climate change from fossil fuels is an example. Nuclear power and the need to find long-term storage for high-level nuclear waste likely fits into this category. Would the case of banning the pesticide DDT be another? Odum (1982) examines the idea in terms of environmental change, pointing out how small individual decisions can significantly alter ecosystems, requiring then great public efforts to reverse the degradation. Odum points to lake eutrophication as an example. We can add how Florida marshlands have become sentinel environments for species change – deformity in frogs, feminization of male alligators – which may be early warnings for humans. From the past, we can note decisions from individual companies with respect to plant location, waste control and the proximity of workers' homes led to high levels of infectious diseases and poor air quality.

These ideas and the applied work we have carried out point to the importance of examining environment and health issues in an interdisciplinary way, with social perspectives not being in any way relegated to serve quantitative studies. The centrality of uncertainty and risk and the political nature of most environmental challenges (especially in so far as they affect health and well-being) have provided us with the rationale for this book. Furthermore, there is a need to codify and set down this social perspective for environment and health students and practitioners.

We want to acknowledge assistance we had putting the book together. Thanks to Karen Van Kerkoerle from Western University's Cartography Section in the Department of Geography. She helped us with several figures, Figures 8.3 and 10.1 in particular. We also want to thank Emily Eyles for proofreading chapters and otherwise helping assemble these pages. We are grateful to publishers who have provided permission to use various figures and tables, as well as agencies and individuals who have provided permission-free images and tables – details are provided in the captions. The people at Ashgate have been extremely helpful and accommodating – Katy Crossan, Margaret Younger and Carolyn Court. We also want to acknowledge the many students and colleagues we have worked with over the years on risk, environment and health projects – it is through collaborating with you that many of the ideas in this book took shape.

John Eyles and Jamie Baxter

References

Kahn, A.E. 1966. The tyranny of small decisions: Market failures, imperfections, and the limits of economics. *Kyklos*, 19, 23–47.

Odum, W.E. 1982. Environmental degradation and the tyranny of small decisions. *BioScience*, 32(9), 728–729.

1 Quantitative environmental health

In this opening chapter, the role of quantitative assessment of environmental health risk will be examined through the lenses of environmental epidemiology, toxicology and exposure science.

The bases of these approaches are quantitative and they inform most risk assessments and therefore the bases of most public health action. They are replete with uncertainty but tend to form the basis for the definition of risk and hazard. But precise meanings of the measurements, the creation of false positives and negatives and the public and political response have not reduced scientific uncertainty and public risk characterization on many issues. There have, however, been different methodological approaches to assist in understanding risk, hazard and exposure. These have adopted a variety of qualitative methods, including depth interviews, stories, linguistic analyses and policy interrogation. Their contribution will be examined.

Introduction

Much quantitative work has been undertaken, especially around human exposure to environmental hazards, to develop standards for regulating the manufacture, use and release of chemicals into the environment. The quantitative data used to develop risk estimates usually come from laboratory animal studies employing relatively high dose levels. But interpolation from high to low dose levels and extrapolation from laboratory animals to humans are necessary. For cancer end points, there is no safe dose or exposure, and regulation is required. For non-cancer end points the no-observed-adverse-effect level (NOAEL) for the critical effect is established, and then safety or uncertainty factors (UFs) to account for scientific uncertainties in the total data base, such as response variability within and between species and the lack of chronic exposure data, are applied. The resulting value is a reference dose (RfD), that is the dose at or below which there is unlikely to be any excess risk. One difficulty with the NOAEL/UF approach is that it does not provide a basis for estimating risk at doses above the RfD; thus, if the exposure assessment indicates that human exposure is above the RfD, statistical and biologically based dose-response modelling is required (see Kimmel, 1990).

In their book on looking at environmental risk, Fjeld et al. (2007) confirm this view and establish that there are four assessments that must be carried out, those relating to release of the material of concern, its transport, population exposure and finally consequences of exposure. This approach can be diagrammatized.

Risk calculation then is key to understanding the level and probability of an adverse impact on human health – the computational heart of a risk analysis. It quantitatively links the characteristics of a hazard to its impact on health.

But why quantitative research? It is often seen as more reliable and objective, can lead to generalizable findings if the sample is representative of the population, can establish relationships between variables to accept or reject a hypothesis and can restructure a complex problem to a limited number of statements. These are useful characteristics, and quantitative processes today have their roots in Comte's positivism, which emphasized the use of the scientific method through observation to empirically test hypotheses, explaining and predicting what, where, why, how, and when phenomena occurred. But the scientific method emerged in the 13th century with the drive to quantify data (see Williams, 2007) among early capitalists and the church and its holdings. Since then quantitative research has dominated Western culture as the research method to create meaning and new knowledge, coming front and centre in the 19th century with measurement being seen as part of human progress (see Zilsel, 1942; Gould, 1981). Leedy and Ormrod (2001) comment that quantitative research is specific in its surveying and experimentation, as it builds upon existing theories, assuming an empiricist paradigm (Creswell, 2003). Within such an approach, the research itself is meant to be independent of the researcher. As a result, data is said to be used to objectively measure reality. Quantitative research creates

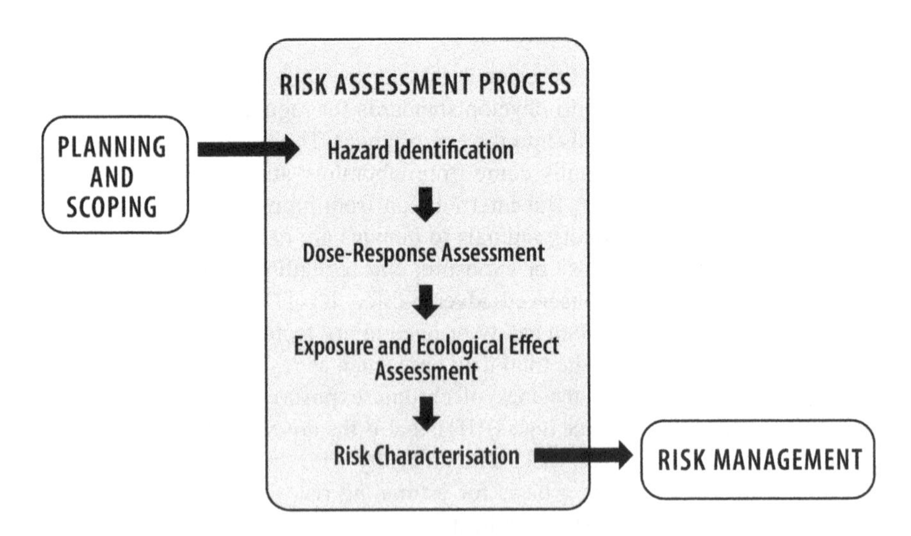

Figure 1.1 The risk assessment and management process

Adapted from EPA (2014).

meaning through objectivity uncovered in the collected data. Quantitative studies thus frame problems as relational questions of variables. "Quantitative researchers seek explanations and predictions that will generate to other persons and places. The intent is to establish, confirm, or validate relationships and to develop generalizations that contribute to theory" (Leedy and Ormrod, 2001, p. 102). Though a major objective of this book is to highlight other, often non-quantitative, approaches to thinking about environment and health it is important to delineate more thoroughly the dominant quantitative approaches to environment and health problems. Toxicology and epidemiology are prominent among the environment and health disciplines that emphasize a quantitative positivist approach to understanding.

Toxicological interventions

These positivist principles seem extremely useful for assessing the relationship between a contaminant and ecological or human health impact. One of the most important sciences in this regard is toxicology, which examines how environmental exposures lead to specific health outcomes, generally in animals, as a means to understand possible health outcomes in humans. Toxicology has the advantage of being able to conduct randomized controlled trials and other experimental studies because animal subjects are used. However, there are many differences in animal and human biology, and there can be uncertainty when interpreting the results of animal studies for their implications for human health. The use of animal studies on environmental cancer investigations will be portrayed and as an example their role in estimating polychlorinated biphenyls (PCBs) and polybrominated diphenyl ethers (PBDEs) exposure.

The complexity of the many cancers and the urgent need for improved patient outcomes provide the impetus to employ animal models for cancer, a common method in toxicology. The most commonly used animal cancer models are rodents – mice and rats. Other cancer models include hamsters, rabbits, dogs, cats, livestock and fish. Toxicologists perform research in whole animals to ensure the short-term and long-term safety of such products as new drugs, cleaning products, plastic food containers, flame-retardant infant clothing and food additives before they are brought to market.

> Toxicity testing in animals is conducted to identify possible adverse effects resulting from exposure to an agent and to develop dose-response relationships that allow evaluation of responses at other exposures. Toxicity tests are designed to minimize variance, bias, and the potential for false-positive and false-negative results.
>
> (NRC, 2006, p. 26)

Acute toxicity tests examine the potential impact of short-term exposure. Historically, an acute toxicity test determined a chemical's median lethal dose (LD50), the dose that causes death in 50% of the test animals. Acute studies can show whether toxicity is sudden, delayed, time limited or continuous. The time to onset and resolution of toxicity can provide insight into the time course of bodily absorption, distribution and clearance of a toxicant. Acute toxicity data can also provide some idea of relative bioavailability. However, extrapolating dose response to humans is not an easy task because of different metabolic rates.

Most toxicological tests using animals use two species. For the animal studies, it is typically one rodent and one non-rodent species. There is some choice in the selection of the non-rodent species because dogs, non-human primates and mini-pigs are available. The extrapolation of animal toxicology data to humans is optimal when the most 'human-like' species is used for the studies. Further, genetic changes can be made to the animals to make them more human-like. For example, toxicological studies of exposure to phthalates in humans arrive at inconclusive results with respect to diabetes or obesity. A particular receptor peroxisome proliferator–activated receptor (PPAR; organotins and phthalates) are intimately involved in the regulation of adipocyte differentiation, production of adipokines, insulin responsiveness and other biological processes related to glucose and lipid regulation (see Thayer et al., 2012). Higher concentrations are required in humans to activate PPAR, so human genes replace the animal ones. Then when rodents are administered with bis(2-ethylhexyl) phthalate (DEHP), a highly lipophilic (fat-soluble) substance used in polyvinyl chloride (PVC) plastic, which can be chemically bonded to this plastic and readily leach into blood or other lipid-containing solutions, the mice gained weight and had increased epididymal white adipose mass compared with wild-type animals. This suggests that concerns over plasticizers have some scientific validity, and animal studies may greatly assist, particularly in this case with the health of infants and babies.

In all toxicological tests, the potential adverse health end points are key. For reproductive and developmental health, there are multiple stages of exposure and variability and a wide array of possible effects. For example the National Research Council (2006, p. 45) suggests,

> Exposures of sexually mature animals can result in sterility or decreased fertility by depleting or affecting ova or sperm or by affecting endocrine functions of organs involved in reproduction. If fertilization occurs, abnormalities of ova and sperm can result in embryonic death, failure of implantation, congenital malformations, embryonic growth retardation, genetic disease, or cancer in the offspring. Exposures during pregnancy can result in embryonic or fetal death, congenital malformations, reversible or irreversible growth retardation, or premature or delayed parturition; they may also have delayed postnatal effects, such as cancer, neurobehavioral effects, growth retardation, and death. Toxicant exposures of neonatal, immature, or adolescent organisms may result in growth retardation or stimulation, endocrine abnormalities, immunologic deficits, neurobehavioral effects, cancer, or death.

This means many tests are required. These include developmental assays, mechanistic studies and the consolidation of various testing results into databases, allowing for the estimation of human developmental toxicity (NRC, 2006).

Risk assessments for end points are typically based on a magnitude of exposure to determine a threshold at or below which effects do not occur and above which they do. But to determine what dose is unsafe is often problematic because of different sensitivities in populations and what may happen at low doses (Moeller, 2009). This means that the dose-response relationship is characterized by values, such as a no-observed-adverse-effect level (NOAEL) or a benchmark dose (BMD; see Figure 1.2). These then serve as the basis for identifying a reference dose (RfD), a reference concentration (RfC), a tolerable intake, or a guidance value – exposure levels at or below which significant adverse effects are thought not to occur. Using the RfD approach requires dose-response data on the critical effect, which is the most sensitive adverse response that occurs at the lowest dose. These dose-response data are used to identify the NOAEL or lowest observable adverse effects level LOAEL (the lowest dose at which there is a significant increase compared with the control group). Conservative measures are used in that correction or uncertainty factors (UF) are built into the calculations.

> UFs are used to account for several specific issues, including interspecies extrapolation (UFA), human intraspecies variability (UFH), extrapolation between subchronic and chronic exposure durations (UFS), extrapolation of a LOAEL to a NOAEL if a LOAEL is used (UFL), and concerns about the quality or breadth of the database (UFDB).
>
> (NRC, 2006, p. 127)

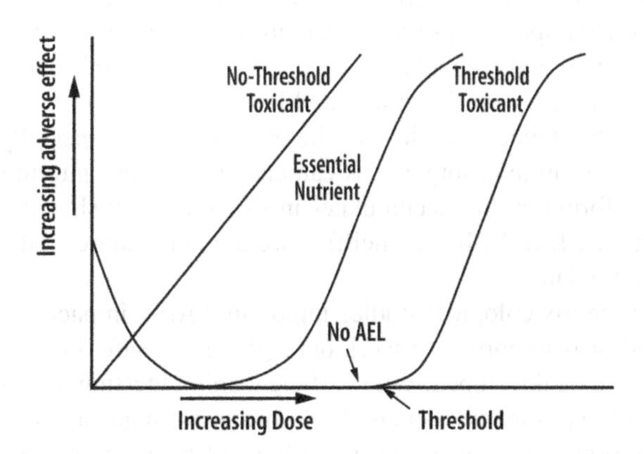

Figure 1.2 Dose-effect curve

Source: Duffus and Worth, IUPAC (2001). Reproduced by permission of International Union of Pure and Applied Chemistry.

For cancer end points, human, animal bioassay and biological indicators are also used. Strong human epidemiological and animal evidence will lead to the conclusion that a substance is carcinogenic. Weak human and strong animal evidence suggests the material is probably carcinogenic for humans. Insufficient evidence leads to the conclusion that it is not possible to determine its carcinogenicity for humans. For BMD, mode of action, type of tumour, target organ and population susceptibility are considered. Though precaution is built into the science vis-à-vis safety factors, the conclusion to draw from 'insufficient evidence' toxicological investigations may lead to different policy decisions depending on the purported benefits of the substances involved.

The necessity of animal studies has been shown in examining methylmercury and polychlorinated biphenyls (PCBs). It is difficult to isolate specific effects without advanced behavioural procedures and controlled exposures using laboratory animals. Studies showed that methylmercury's profile is dominated by sensory effects with a likely action in the brain at the cerebral cortex. Some of these effects may be amplified with aging. Its effects are cognitive, affecting environmental appraisal but not memory. For example, methylmercury and some PCBs can disturb human response times to environmental stimuli (see Newland and Paletz, 2000). As the EPA (2013) noted, PCBs have been demonstrated to cause a variety of adverse health effects, including cancer in animals. They have also been shown to cause a number of serious non-cancer health effects in animals, including effects on the immune system, reproductive system, nervous system, endocrine system and other health effects. Studies in humans provide supportive evidence for potential carcinogenic and non-carcinogenic effects of PCBs (Carpenter, 2006). PCBs were common between 1929 and their North American banning in 1979. Due to their non-flammability, chemical stability, high boiling point and electrical insulating properties, PCBs were used in hundreds of industrial and commercial applications including electrical, heat transfer and hydraulic equipment; as plasticizers in paints, plastics and rubber products; in pigments, dyes and carbonless copy paper; and many other industrial applications. Methylmercury on the other hand is a by-product of mercury, a naturally occurring metal. Once deposited in the air and water, especially after burning fossil fuels, certain microorganisms can change mercury into methylmercury, a highly toxic form that bio-accumulates in fish and shellfish and biomagnifies in animals that eat fish. Fish and shellfish are the main sources of methylmercury exposure to humans.

Why are the toxicological studies important? After impacts on human health are detected, it is important to work out if there is a safe dose and if alternative products could be developed. With sufficient animal testing, toxicology becomes a vital part of exposure assessment. Exposure assessment calculates human exposure to environmental contaminants by both identifying and quantifying exposures. Exposure science can be used to support environmental epidemiology by better describing environmental exposures that may lead to a particular health outcome or identify common exposures whose health outcomes may be better understood through a toxicology study, or they can be used in a risk assessment to determine whether current levels of exposure might exceed recommended levels. Exposure science has the advantage of being able to very accurately quantify exposures to specific chemicals, but it does not generate any information about health outcomes like environmental epidemiology or toxicology. In exposure science, quantitative measurements provide a direct link between exposure and the dynamic processes of contact and the kinetic processes that define the levels or the form of a substance after it has been metabolized and redistributed, bio-accumulated or removed from the body. Choice and development stage of laboratory animal selection is then vitally important (EMA, 1995). The use of biomonitoring in exposure science is also important, because the results can identify highly exposed or unexposed individuals when compared with a population distribution. Whereas toxicology studies non-human animals, human populations are the focus of epidemiologic research.

Environmental epidemiology

Environmental epidemiology employs outputs from toxicology and exposure science to understand the relationships between environmental exposures (including exposure to chemicals, radiation, microbiological agents, etc.) and human health. Observational studies, which simply observe exposures that people have already experienced, are common in environmental epidemiology because humans cannot ethically be exposed to agents that are known or suspected to cause disease. Nevertheless, we have already pointed out that toxicological analysis can be flawed despite best efforts to build in safety factors to account for analytical uncertainty. There are, for example, chemicals that slip through because the human health outcome was not anticipated. Thalidomide, a drug once used to control symptoms of morning sickness in pregnant women, was eventually discovered through epidemiological analysis to lead to deformities in the children exposed during pregnancy.

While the inability to use experimental study designs is a limitation of environmental epidemiology, this discipline has the advantage of direct observation of effects on human health rather than estimating effects from animal studies. Environmental epidemiology is then the study of the determinants of the distributions of health and disease outcomes in human populations (see Hertz-Piccioto, 1998). Exposure is often involuntary (e.g., exposure to substances released by a nearby facility) but may also be voluntary for substances approved for use under certain conditions (e.g., exposure to pesticides from appropriate or inappropriate application). Different research designs are used to associate environmental exposure with health outcomes, including cross-sectional designs involving observation of all of a population or a representative subset, at one specific point in time; ecological, which investigates aggregate populations rather than individuals; case-control for comparing subjects who have that condition/disease with patients/residents who do not have the condition/disease but are otherwise similar; and cohort studies, which are largely about the life histories of segments of populations and the individual people who constitute these segments. There are modifications of such

Table 1.1 Observational study designs in epidemiology

Cohort studies	A cohort (group) of individuals with exposure to a chemical and a cohort without exposure are followed over time to compare disease occurrence.
Case-control studies	Individuals with a disease (e.g., cancer) are compared with similar individuals without the disease to determine if there is an association of the disease with prior exposure to an agent.
Cross-sectional studies	The prevalence of a disease or clinical parameter among one or more exposed groups is studied, for example, the prevalence of respiratory conditions among furniture makers.
Ecological studies	The incidence of a disease in one geographical area is compared to that of another area, for example, cancer mortality in areas with hazardous waste as compared to areas without waste sites.

Source: Encyclopaedia of Earth (2013).

Table 1.2 Types of potential error environment and health studies

Selection bias	Occurs when the study group is not representative of the population from which it came.
Information bias	Occurs when study subjects are misclassified as to disease or exposure status. Recall bias occurs when individuals are asked to remember exposures or conditions that existed years before.
Confounding factors	Occur when the study and control populations differ with respect to factors which might influence the occurrence of the disease. For example, smoking might be a confounding factor and should be considered when designing studies.

methods, such as repeat cross-sectional, which attempts to examine the same or similar individuals with the same exposure at several time periods. In environmental cases, experiments are only possible if exposures can be studied as they occur – a natural or quasi-experimental design. The strongest in terms of producing valid evidence is the cohort study, which can be prospective (cohorts identified of current exposures and followed into the future) or retrospective (cohorts identified on past exposures and then followed from that point).

Such investigations are often quite difficult and seldom offer definitive results because of various types of bias (Table 1.2). They try to establish the significance of their observations by measures of effect, such as odds ratio (ratio of risk of adverse health outcome in an exposed population compared with a similar non-exposed population) and relative risk (ratio of elevated risk in an exposed group compared with a reference population; see for example Schmidt and Kohlmann, 2008; Figure 1.3).

An example of a prospective cohort study is an examination of independent associations between childhood exposures to smoking and household dampness and phlegm and cough in adulthood by Cable et al. (2014). This study utilized 7,320 members of a British cohort who were born during one week in 1970 for whom there was complete childhood and adult information. Cable et al. (2014) used measures pertaining to coughing and phlegm among participants when they reached 29 years of age. Exposure to smoking and household dampness at age 10 were both associated with a higher relative risk for phlegm with those living in damp conditions being twice as likely to report this outcome. Cough was related by exposure to adult smokers in the home at age 29; current smoking at age 29 contributed to cough and phlegm incidence; and a substantial association between household dampness and co-occurring phlegm and cough suggests long-term detrimental effects of childhood environmental exposures. Their findings give support to current public health interventions for adult smoking and raise concerns

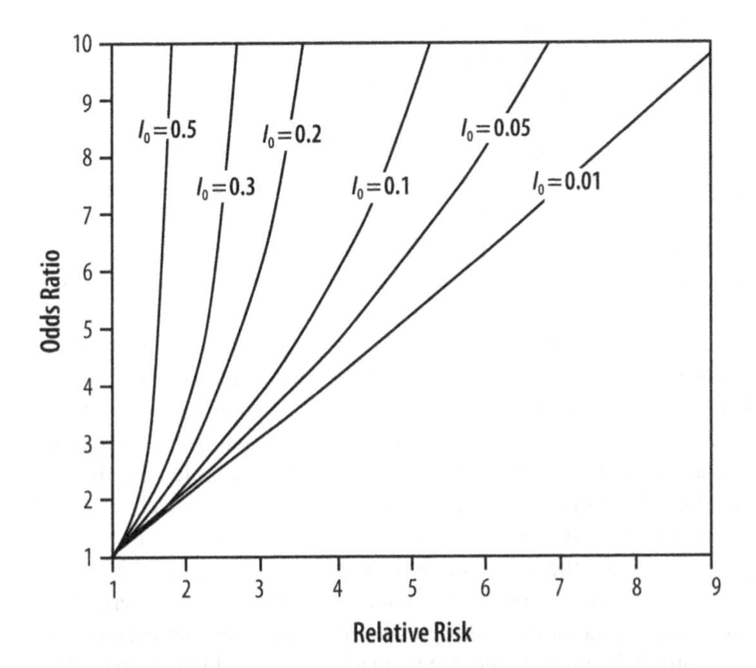

Figure 1.3 Relationship between odds ratio and relative risk for various incidence rates

Source: Schmidt and Kohlmann (2008), with kind permission from Springer Science and Business Media.

about the long-term effects of a damp home environment on the respiratory health of children (Cable et al., 2014).

An example of a retrospective cohort study is provided by the Atom Bomb Casualty Commission (ABCC), which initiated the life-span study of the survivors of the atomic bomb explosions in Japan. Exposed survivors were classified according to their distance from the hypo-centre at the time of the explosion and the presence or absence of acute symptoms attributable to irradiation. Rapid conclusive evidence was found that irradiation increased the risk of leukaemia, cataracts and mental retardation in children heavily exposed in utero. Because of many deaths in this cohort, the population of interest was redefined in 1950 and has been followed to the present. By the late 1980s nearly 7,000 deaths from cancer and more than 8,500 registered cases had been recorded in a population of some 76,000. The results have shown a lifetime risk of fatal cancer of about 10% per Sievert,[1] with about a tenth of the deaths attributable to leukaemia.

Cohort studies have been used in circumstances in which data are difficult to obtain at the individual levels, as in the possible association between liver and kidney cancer and non- Hodgkin's lymphoma and TCE, or trichloroethylene, a volatile organic chemical (VOC) used primarily as an industrial solvent, with its most common use to remove grease from fabricated metal parts and some textiles (see EPA, 2007). TCE has been found in ambient air, surface water and ground water. TCE levels above background have been found in industrial settings, in homes undergoing renovation and in homes using private wells located near TCE disposal or contamination sites. The most likely TCE exposure route for children is ingestion of contaminated drinking water. An early occupational cohort study helped set standards for TCE exposure (see Tola et al., 1980). Wartenberg et al. (2000) point to strength of evidence that comes from long, occupationally based cohort studies but also comment that many of these rely on job titles to classify individuals, thereby assuming all individuals with the same job title have a similar exposure to TCE. The studies only occasionally take account of confounding or modifying factors (e.g., smoking, alcohol consumption) and rarely demonstrate an elevated relative risk for those exposed. Yet many community-based studies of TCE determine exposure by place of residence or water supply. There is again limited or no information on possible confounding variables. In general, these are cross-sectional studies (see what follows) of cancers, often childhood cancers, and drinking water contamination. Although dependent on sample size, these studies can have relatively high statistical power (i.e., the ability to detect an effect if one exists) even though exposure levels are relatively low because of the large number of subjects consuming the TCE–contaminated water. The statistical power drops dramatically, however, if rarer health outcomes are chosen like certain cancers. This creates a dilemma for study design since the outcomes of most concern to the public may simply be unstudiable due to a lack of exposed individuals. But exposures typically are to multiple solvents, making it difficult to attribute observed results to only one agent. Exposure is usually assessed at a community level rather than the individual level, requiring assumptions be made about water consumed, impact of population mobility and so on. These investigations do excite intense local interest as in the exploration of the childhood leukaemia cluster and TCE exposure in Woburn, Mass., where no statistically significant relationships have been found and the case reports have technically been unsubstantiated despite years of legal activity (see Lagakos et al., 1986).

According to environmental epidemiologists, case-control studies are strong but not as strong as well-designed cohort studies. As an example, we use an examination of cryptosporidiosis, a common gastrointestinal disease widespread in many developed and developing countries. While much of the impact of this gastrointestinal (GI) infection comes from outbreak studies, a number of case-control studies have been undertaken. These have highlighted risks from drinking water from poorly treated public and private supplies, swimming in pools, and contact with farm animals. Contact with young children is also known to be a risk factor. Lake et al. (2007) studied the role of a wide array of environmental and socioeconomic factors on human cryptosporidiosis whereby 3,368 laboratory-confirmed cases were compared to the locations of an equal number of controls. When all cryptosporidiosis cases were analyzed, several associations emerged, with illness found in areas with a relatively higher proportion of high-socioeconomic-status individuals, higher numbers of individuals aged less than 4 years, areas with land treated with manure and areas with poorer water treatment.

Nested case-control studies present case-control work conducted within a cohort study. Within a cohort study, individuals with a specific disease outcome are compared with controls also from the cohort but without the outcome of interest (see Ernster, 1994). One example is the case of bisphenol A exposure and premature mortality, which Cantonwine et al. (2010) examined in Mexico City. In this study, 670 women agreed to participate in a cohort study. From this group, archived third-trimester urine samples, collected during 2001 to 2003, were available for 518 non-smokers. These formed the basis of the nested case-control study. Thirty participants were selected randomly among women who had completed 38 or more weeks of gestation at the time of delivery. Additionally, 30 participants were selected among women who had delivered before 38 weeks of gestation at time of delivery, which included 12 women who delivered prior to 37 weeks. These samples were then explored for BPA, as it has been shown to weaken oestrogenic activity and potentially impacts thyroid function. Additionally, recent evidence from in vitro studies suggests that foetal exposure to BPA can occur through placental exchange. Furthermore, in animal studies, environmentally relevant concentrations of BPA have been associated with low birth weight. BPA was detected in 80.0% ($N = 48$) of the urine samples; total concentrations ranged from < 0.4 μg/L to 6.7 μg/L; uncorrected geometric mean was 1.52 μg/L. The adjusted odds ratio of delivering less than or equal to 37 weeks in relation to specific gravity adjusted third-trimester BPA concentration was 1.91 (95% CI 0.93, 3.91, p-value = 0.08). When cases were further restricted to births occurring prior to the 37th week ($n = 12$), the odds ratio for specific-gravity–adjusted BPA was larger and statistically significant ($p < 0.05$). This investigation has shown a possible association between BPA exposure and premature births, albeit in a small sample (Cantonwine et al., 2010).

The most commonly used design is the cross-sectional study, carried out at one time point or over a short period. They are usually conducted to estimate the prevalence of the outcome of interest for a given population, commonly for the purposes of public health planning. Data can also be collected on individual characteristics, including exposure to risk factors, alongside information about the outcome. In this way cross-sectional studies provide a 'snapshot' of the outcome and the characteristics associated with it. Cross-sectional studies are sometimes carried out to investigate associations between risk factors and the outcome of interest. But they give no indication of the sequence of events – whether exposure occurred before, after or during the onset of the disease outcome. This being so, it is impossible to infer causality. The level of nonresponse is another concern, but a greater one still is that of biased response, in which a person is more likely to respond when they have a particular characteristic or set of characteristics. Bias will occur when the characteristic in question is in some way related to the probability of having the outcome (Levin, 2006). Shepherd et al. (2014) carry out a cross-sectional study comparing noise annoyance and health-related quality of life of individuals residing close to a major international airport or wind turbine complex with those located in demographically matched areas. Noise is an environmental nuisance that has the potential to degrade health and negatively impact the relationship between humans and their environments. It was identified as a significant environmental health challenge as early as the 1970s. At high intensities, noise can induce hearing loss. At low levels, it can significantly impact health by interfering with sleep or inducing maladaptive emotional responses. Shepherd et al. (2014) found that health-related quality of life may be degraded in those living in areas more likely to produce noise annoyance. Furthermore, the addition of aviation noise to environments already affected by road noise may induce further annoyance and degradations in health-related quality of life, indicating that one noise source may not mask the impact of another (Shepherd et al., 2014). This study in fact points to a common use of cross-sectional surveys to identify psycho-social health effects such as odour annoyance in rural areas (see Blanes-Vidal et al., 2012) or anticipatory fears around land-use changes (see Ochodo et al., 2014). The problem is that responses to questions meant to measure health outcomes in a survey may be conditioned by local controversy and debate. Yet for some outcomes (e.g., feeling rested, annoyance) there may not be suitable 'objective' alternatives, as is arguably the case when detecting something like stress through chemical changes in urine or hair or sleep patterns through a brain activity electroencephalogram (EEG).

Repeated cross-sectional studies can add greater explanatory power. Several were carried out after German reunification. Sugiri et al. (2006) observed an improvement in lung function with decreasing levels of total suspended particulates (TSP) and SO_2 among 2,574 East German 6-year-old children, equivalent to their western counterparts by the time of the third survey, 8 years after reunification. The improvement was weaker in children living within 50 m of a busy street – a finding that was attributed to the 50% to 75% increase in motor vehicles during this period in eastern Germany. Frye et al. (2003) reported improvements in lung capacity over 3 consecutive cross-sectional surveys (1992–1997) of 2,493 11- to 14-year-old children in three East German communities. Forced vital capacity (FVC) of lungs increased by 4.7% for a 50-g/m^3 decrease of TSP and 4.9% for a 100-g/m^3 decrease of SO_2, whereas effects for FEV_1 were smaller (see Götschi et al., 2008).

A geographical ecological design was used to explore Q fever (a zoonotic disease resulting from inhalation of cell variants or contact with fluids of infected animals) in the Netherlands. As Reedijk et al. (2013) note this approach can be useful for decision making because health policy is usually conducted at the municipal, regional or national level rather than the individual level. They also comment that the influence of environmental factors is difficult to assess at the individual level. But a major limitation of ecological design is that no causal inference can be made between exposures (risk factors) and outcome (Q fever) at the individual level. Furthermore, some potentially important risk factors were not included as they were not available at postal code level.

Natural experiments are studies that leverage events, interventions or policies which are not under the control of researchers but can be amenable to research using the variation in exposure that they generate to analyze impact. Examples include the effect of famine on the subsequent health of children exposed in utero or the effects of clean-air legislation, indoor smoking bans and changes in taxation of alcohol and tobacco (MRC, 2009). One example paralleled a longitudinal study of the development of psychiatric disorder to learn the impact of increased household income. This began in 1993 and took advantage of the establishment of a casino on an American Indian reservation in the study area. Approximately one quarter of the children in the study belonged to families that received a royalty income from the casino's profits from 1996 onwards. Costello et al. (2003, 2010) used data from the survey to classify families as persistently poor, ex-poor or never poor in order to assess the effect on children's mental health of an increase in household income sufficient to lift them out of poverty. In the children whose families moved out of poverty, the overall frequency of psychiatric symptoms fell to levels similar to those of the never-poor children. It remained high in the children whose families remained poor. Costello et al. (2003, 2010) concluded that findings suggest that social determinants of health rather than individual predispositions explained the link between child poverty and psychopathology. All the Indian families received the casino income, largely breaking the link between movement out of poverty and family characteristics that might influence behavioural symptoms in the children. A similar pattern in the non-Indian children was observed. The researchers used the good fortune of the casino opening to conduct an additional study, a natural experiment, which could use the variables already measured. A further follow-up in 2006, when the youngest cohort members were aged 21, showed that the protective effects of the extra income persisted into early adulthood (see Costello et al., 2003, 2010).

True experimentation requires that the experimenter assign 'subjects' to exposure and control groups. Thus, natural experimentation depends on carefully matching the exposed group with an unexposed group that is similar in every other way. This allows environmental epidemiologists to isolate exposures as causes by controlling for confounding variables (e.g., smoking, diet, job status). Overall strength of evidence claims made by epidemiologists are associated with the postulates developed by Bradford Hill (1965; see Table 1.3). These criteria have been much used and debated as a rank-order list of conditions that, if met, bolster the idea that an association, for example between an environmental exposure and a negative health outcome, is a true causal effect.

Table 1.3 Hill's criteria for causation

Strength	The stronger the association, the farther from zero the measure the effect studied will be. Consider that strong associations are more likely to be causal.
Consistency	By repeating findings in different populations, similar outcomes may be found.
Specificity	The presence of the cause is necessary for the effect to happen.
Temporality	Exposure always precedes the outcome.
Biological gradient	The dose-response phenomenon happens. Therefore, and increasing level of exposure (in amount and/or time) increases the risk.
Plausibility	If the relationship that is being studied is plausible based on existing biological knowledge, there is more chance that the observed relationship may be causal.
Coherence	The causal interpretation must not conflict with existing theory or knowledge at the time of the investigation.
Experiment	Causal relationships are best demonstrated by experimental evidence, specifically that the condition can be altered (prevented or ameliorated) by an appropriate experimental regimen.
Analogy	Finding analogous associations between similar factors and similar diseases. This type of analogy raises the chance that the observed association may be causal.

Source: Adapted from Bradford Hill (1965).

For example, Pruss (1998) used the criteria to review 22 studies which suggest that there is a causal relationship between the gastrointestinal symptoms and recreational water quality, measured by indicator-bacteria concentration. They report a strong and consistent association with temporality and dose-response relationships, as well as biological plausibility and analogy to clinical cases in drinking water pollution. Yet from a policy perspective we must be careful about not acting in the absence of satisfying many criteria on this list. Epidemiologists are cautious about identifying causation because it is a tenet of science to be conservative about accepting hypotheses as 'true'. However, what his good for science may not always be in the best interest of those concerned to protect public health. For example, setting statistical significance at 0.1 rather than 0.01 may be more precautionary. But Bradford Hill was ambivalent about causation and considered public health action more important than determining cause if strong associations exist. "Causal inference in epidemiology is better viewed as an exercise in measurement of an effect rather than as criterion-guided process for deciding whether an effect is present or not" (Rothman and Greenland, 2005). Yet it may be argued that strong associations on all the criteria suggested by Hill will be influenced by study design so that cohort and case control are more likely to support strong associations than cross-sectional and ecological designs because the former two satisfy several criteria by virtue of the design itself (e.g., cohort satisfies temporality). Even then as some of our examples have shown, demonstrating a clear relationship between exposure and outcome is potentially shaped by confounding and modifying factors as well as recognition and control of bias.

So why social perspectives? The bases of these approaches are quantitative, and they form the foundations of most risk assessments and therefore the core of much public health action. As we have seen they are replete with uncertainty, but because they represent the best scientific tools available, they tend to form the basis for defining levels of risk and action on technological, pollution and contamination hazards. But precise meanings of the measurements and the knowledge of the risk of false positives and negatives can get forgotten in the rush to form public and political responses to hazards. Risk characterization is sometimes highly imprecise – for example, for new substances – so policy makers and public health professionals face the challenge of sifting through the sources of uncertainty to best ensure public safety. This is especially the case when it is necessary to respond quickly to public concerns. But even with sufficient time and close attention to design specification, environmental epidemiology is necessary but limited unless experimentation and biological plausibility can be invoked. As Oreskes (1998, p. 1453) argues with respect to lead exposure, "even with this contaminant it is not possible to demonstrate the predictive reliability of any model of a complex natural system. All models embed uncertainties, and these uncertainties can and frequently do undermine predictive reliability." In the case of lead in the environment, we may categorize model uncertainties as theoretical, empirical, parametrical and temporal. Theoretical uncertainties are aspects of the system that are not fully understood, such as the biokinetic pathways of lead metabolism. Empirical uncertainties are aspects of the system that are difficult (or impossible) to measure, such as actual lead ingestion by an individual child. Parametrical uncertainties arise when complexities in the system are simplified to provide manageable model input, such as representing longitudinal lead exposure by cross-sectional measurements of likely exposure pathways. Temporal uncertainties arise from the assumption that systems are stable in time. A model may also be conceptually flawed. The Ptolemaic system of astronomy is a historical example of a model that was empirically adequate but based on a 'wrong', that is physically false Oreskes (1998). The author is questioning the scientific models that underpin much environmental epidemiology. It is hard to escape or even consider these limitations. Measurement has been significantly improved over the past 20 years or so and new ways of identifying exposures, outcomes and their linkages identified. But their applicability in understanding risk and hazard is not always clear. For example, Franco et al. (2004) comment that the use of most biomarkers is limited as they usually measure only relatively recent exposure, whereas most cancers, for example, take decades to develop. Most studies still rely on the difficulty of controlled observations of exposures and outcomes, and there can be little attention given to societal forces in shaping risk at this individual level of analysis. For example, what would a simple measure of community impact on individuals look like?

There have, however, been different methodological approaches to assist in understanding risk, hazard and exposure. These have adopted a variety of qualitative methods, including depth interviews, stories, linguistic analyses and policy interrogation, which form an important base for social perspectives. These are not meant to supplant toxicological, epidemiological or risk analysis approaches, but instead they can expand our knowledge in areas where these sciences leave gaps. The use of interviews has many things in common with cross-sectional surveys (part of both quantitative and qualitative approaches) but may be more appropriate in the context of risk controversies involving immense local and political scrutiny. For example, Williams et al. (2003) review community studies of the impact of racial and ethnic discrimination on health status and do find associations, especially with mental health. But they comment that it is difficult to attribute adverse health with discrimination. Three case studies of technological disasters in the United States – a train derailment and toxic spill in Livingston, Louisiana; a neighbourhood contamination by a superfund site in Houston, Texas; and an 11-million-gallon oil spill in Prince William Sound, Alaska (see Gill and Picou, 1998) find high levels of chronic community stress and a relationship between perceived threat to health and level of community stress. Lambert et al. (2006) report on a survey of residents in Sydney, Nova Scotia, and their perceptions of such problems as smoke, soot, ash deposition and perceived dangers to children. In what may be seen as a nested approach, Bush et al. (2001) selected a subset of survey respondents for follow-up to identify dimensions of stigma and psycho-social stress in Teesside. Finally, Venables et al. (2009) questioned people about new nuclear building programs and establish four landscapes of belief – Beneficial and Safe; Threat and Distrust; Reluctant Acceptance; and There's No Point Worrying. Views about nuclear power in such communities are both subtle and complex, avoiding simplistic bipolar dichotomies such as 'for' or 'against', and that there is a need for extensive and meaningful dialogue with such communities over any new build plan. All these surveys and many more point to the immediate political response necessary from such surveys, even if they might be regarded as scientifically unsound epidemiologically.

Unlike epidemiology, which often relies on large sample sizes, alternate approaches may do much with smaller sample sizes by trading that off against going for greater depth of reports about health-related experiences in environmental challenges. Sometimes in-depth interviews have been carried out with key stakeholders, particularly to interrogate a policy direction or with citizens to explore experiences. With respect to the first, Edge and Eyles (2013) examined the viewpoints of government, business, science and non-governmental organizations on the banning of bisphenol A from baby bottles in Canada under the Chemical Management Plan. Indeed once the government had decided on the ban all fell into place, especially as the economic costs of the ban were small. More contentious were the debates about the chemical management plan especially with respect to the relationship between scientific evidence and intervention. Questions about what evidence, collected by whom and what exposure should be treated as a risk, given the pro-business, trade regulation–impacted role of government, were contentious. Many assessments of chemicals were put on the side, to be re-visited when new evidence became available. In the interim, the status quo – in some cases using the substance – was seen as an acceptable stance.

The views of citizens impacted by environmental events have been frequently explored using depth interviews with a selected number of people, often sampled by distance from the event, or family status or interest in the issue. It is recognized that such sampling is partial, but identifying themes until no others emerge (saturation) is a legitimate method (Morse, 1995), in this case for conceptualizing the social/contextual aspects of risk views. Since the early 1980s, Edelstein (2003) has been obtaining citizen views and experiences of polluted situations, especially in Legler, New Jersey. The presence of toxic exposure can undermine the very fabric of society. It leads to a loss of trust, the inversion of the home (formerly seen as a safe haven, now hopelessly poisoned), a sense of a loss of control in one's personal life and, over the present and the future, a different relationship to and assessment of the environment (now seen as dangerous, and insidious in its dangers) and a pessimistic attitude towards health. It places the adults in contaminated families under a great deal of stress as they become isolated and stigmatized by their contamination, and it teaches children to fear.

Edelstein used interviews then to develop conceptual insights about fear, trust and stigma which have been widely absorbed in the literature. Lifescape change is being posited. At roughly the same time, in the mid-1970s Love Canal, New York, became the subject of national and international attention after it was revealed that the site had formerly been used to bury 22,000 tons of toxic waste by Hooker Chemical Company (now Occidental Petroleum Corporation). Hooker Chemical sold the site to the Niagara Falls School Board in 1953 for $1, with a deed explicitly detailing the presence of the waste, with a liability limitation clause about the contamination. The construction efforts of housing development, combined with particularly heavy rainstorms, released the chemical waste, leading to a public health emergency and an urban planning scandal. What was there? Beck (1979) reported at the time that, "twenty five years after the Hooker Chemical Company stopped using the Love Canal as an industrial dump, 82 different compounds, 11 of them suspected carcinogens, have been percolating upward through the soil, their drum containers rotting and leaching their contents into the backyards and basements of 100 homes and a public school built on the banks of the canal." The issues of pollution and health impacts remain, but small numbers of cases and of those remaining available to be interviewed means little gets finally resolved (see Gensburg et al., 2009). But Love Canal brought citizens into science (Gibbs, 1982), leading to greater government involvement and regulation and sometimes conflict with business. This is also shown by the contamination of ground water at Woburn, Mass., and what was seen as a cluster of childhood leukemia cases (see Brown, 1987, 1992). Use of citizen eye-witness accounts thus formed the basis of popular epidemiology and the centrality of lay perspectives in environmental health matters.

Another method to clarify the social perspective is the analysis of policy documents often through examining how language is used. For example, Garvin and Eyles (2001) use such an approach to examine national differences in public health policies using a case study of Sun Safety programs in Australia, Canada and England. A single public health issue identified at the global scale (rising skin cancer rates) is framed differently based upon specific social, cultural and political situations. This results in a different story, or narrative, embedded in each national policy, identified by national documents and policy initiatives. Policy document analysis can also note what is not part of risk evidence/expertise, as in the case of health concerns about fracking in some states in the United States. The Marcellus Shale is a vast natural gas field underlying parts of Pennsylvania, New York, West Virginia, Virginia and Maryland. Rapid development of this field has been enabled by advances in hydrofracking techniques whereby water displaces the gas, but it has raised concerns about contaminated leaching of this water. Response to public concern about potential adverse environmental and health impacts has led to the formation of state and national advisory committees. Although public health is specified to be a concern in the executive orders forming three advisory committees, no individuals with health expertise among the 52 members of the Pennsylvania Governor's Marcellus Shale Advisory Commission, the Maryland Marcellus Shale Safe Drilling Initiative Advisory Commission or the Secretary of Energy Advisory Board (SEAB) Natural Gas Subcommittee could be found (see Goldstein et al., 2012).

It is common to combine depth interviews with document analysis. As Bernier and Clavier (2011) note policy analysis is more than document analysis and involves interviews, conceptual lenses and the recognition that policy making is political. We turn in the next chapter to revealing some of these dimensions. It is important in using qualitative approaches, therefore, to identify the big as well as the small stories (see Georgakopoulou, 2006). Small stories provide the experiences and insights of citizens and stakeholders involved in an event. A big story makes sense of that conceptually either as citizen or analyst (see Caru and Cova, 2008). These authors carried out such an approach with respect to consumption experiences. Although in itself not framed in small/big story terms, the story of plastics and their production and use can be treated as many small stories of beneficial in their use and provision of community employment. Big stories can be seen as corporate greed, government inaction, environmental despoliation and health damage (see Freinkel, 2011).

Quantitative toxicological, epidemiological approaches within risk analysis were meant for a particular purpose – to decide the safe levels of substances – but these approaches leave gaps. They do not account for social values and the politics of decision making on risk. Coalitions of concerned citizens may as easily set aside the uncertainties and limitations of these quantitative approaches as they emphasize these uncertainties and limitations in arguing for policy change on technological hazard risk. At the end of the day, we must recognize that these quantitative sciences will only form part of the decision-making process for risk at the societal level (Sabatier, 1987; Kasperson, 1988). Further, impacts themselves can be social in that debate on risk can cause such psychosocial impacts as upheaval and community conflict. These are all entry points for the application of social scientific methods to expand our understanding of how individuals and groups experience and take action risk in their everyday lives. These unfold in the ensuing chapters.

Note

1 A Sievert (Sv) is a derived SI (International System of Units) unit of ionizing radiation dose, which is to say, a measure of the health effects of low doses of ionizing radiation. 1 Sv is equivalent to the maximum lifetime dose allowable for NASA astronauts. 1 Sv is also equivalent to 100 rem, a similar measure used in the United States.

References

Beck, E. 1979. The Love Canal tragedy. *EPA Journal*, 5(1), 17–20.

Bernier, N.F., & Clavier, C. 2011. Public health policy research: Making the case for a political science approach. *Health Promotion International*, 26(1), 109–116.

Blanes-Vidal, V., Nadimi, E.S., Ellermann, T., Andersen, H.V., & Løfstrøm, P. 2012. Perceived annoyance from environmental odors and association with atmospheric ammonia levels in non-urban residential communities: A cross-sectional study. *Environmental Health*, 11, 27.

Brown, P. 1987. Popular epidemiology: Community response to toxic waste-induced disease in Woburn, Massachusetts. *Science, Technology, and Human Values*, 12(3–4), 78–85.

Brown, P. 1992. Popular epidemiology and toxic waste contamination: Lay and professional ways of knowing. *Journal of Health and Social Behavior*, 33, 267–281.

Bush, J., Moffatt, S., & Dunn, C. 2001. 'Even the birds round here cough': Stigma, air pollution and health in Teesside. *Health & Place*, 7(1), 47–56.

Cable, N., Kelly, Y., Bartley, M., Sato, Y., & Sacker, A. 2014. Critical role of smoking and household dampness during childhood for adult phlegm and cough: A research example from a prospective cohort study in Great Britain. *BMJ Open*, 4(4), n.p.

Cantonwine, D., Meeker, J.D., Hu, H., Sánchez, B.N., Lamadrid-Figueroa, H., Mercado-García, A. . . . & Téllez-Rojo, M.M. 2010. Bisphenol A exposure in Mexico City and risk of prematurity: A pilot nested case control study. *Environmental Health*, 9, 62. doi: 10.1186/1476-069X-9-62

Carpenter, D.O. 2006. Polychlorinated biphenyls (PCBs): Routes of exposure and effects on human health. *Reviews on Environmental Health*, 21(1), 1–24.

Caru, A., & Cova, B. 2008. Small versus big stories in framing consumption experiences. *Qualitative Market Research: An International Journal*, 11(2), 166–176.

Costello, E.J., Compton, S.N., Keeler, G., & Angold, A. 2003. Relationships between poverty and psychopathology: A natural experiment. *Journal of the American Medical Association*, 290(15), 2023–2029.

Costello, E.J., Erkanli, A., Copeland, W., & Angold, A. 2010. Association of family income supplements in adolescence with development of psychiatric and substance use disorders in adulthood among an American Indian population. *Journal of the American Medical Association*, 303(19), 1954–1960.

Creswell, J. 2003. *Research Design: Qualitative, Quantitative and Mixed Methods Approaches*. (2nd ed.) Thousand Oaks, CA: SAGE Publications.

Duffus, J., & Worth, F. 2001. *Essential Toxicology for Chemists*. Research Triangle, North Carolina: IUPAC.

Edelstein, M. 2003. *Contaminated Communities*. Boulder: Westview.

Edge, S., & Eyles, J. 2013. Message in a bottle: Claims disputes and the reconciliation of precaution and weight-of-evidence in the regulation of risks from bisphenol A in Canada. *Health, Risk & Society*, 15, 432–448.

Encyclopaedia of Earth. 2013. *Toxicity Testing Methods*. Available at: http://www.eoearth.org/view/article/156673/ (Accessed 5 June 2015).
Environmental Protection Agency (EPA). 2007. *Trichloroethylene*. Available at: http://www.epa.gov/teach/chem_summ/TCE_summary.pdf (Accessed 3 June 2015).
Environmental Protection Agency (EPA). 2013. *Health Effects of PCBs*. Available at: http://www.epa.gov/wastes/hazard/tsd/pcbs/pubs/effects.htm (Accessed 3 June 2015).
Environmental Protection Agency (EPA). 2014. *Basic Information*. Available at: http://www.epa.gov/risk/basicinformation.htm (Accessed 3 June 2015).
Ernster, V.L. 1994. Nested case-control studies. *Preventative Medicine*, 23(5), 587–590.
European Medicines Agency (EMA). 1995. *Note for Guidance on Toxicokinetics: A Guidance for Assessing Systemic Exposure in Toxicology Studies*. London: EMA.
Fjeld, R., Eisenberg, N., & Compton, K. 2007. *Quantitative Environmental Risk Assessment for Human Health*. Hoboken: Wiley.
Franco, E.L., Correa, P., Santella, R.M., Wu, X., Goodman, S.N., & Petersen, G.M. 2004. Role and limitations of epidemiology in establishing a causal association. *Seminars in Cancer Biology*, 14, 413–426.
Freinkel, S. 2011. *Plastic*. Boston: Houghton Mifflin Harcourt.
Frye, C., Hoelscher, B., Cyrys, J., Wjst, M., Wichmann, H.E., & Heinrich, J. 2003. Association of lung function with declining ambient air pollution. *Environmental Health Perspectives*, 111, 383–387.
Garvin, T. & Eyles, J. 2001. Public health responses for skin cancer prevention: The policy framing of Sun Safety in Australia, Canada and England. *Social Science & Medicine*, 53(9), 1175–1189.
Gensburg, L.J., Pantea, C., Kielb, C., Fitzgerald, E., Stark, A., & Kim, N. 2009. Cancer incidence among former Love Canal residents. *Environmental Health Perspectives*, 117(8), 1265–1271.
Georgakopoulou, A. 2006. Thinking big with small stories in narrative and identity analysis. *Narrative Inquiry*, 16(1), 122–130.
Gibbs, L.M. 1982. *Love Canal: My Story*. New York: SUNY Press.
Gill, D.A., & Picou, S. 1998. Technological disaster and chronic community stress. *Society & Natural Resources: An International Journal*, 11(8), 795–815.
Goldstein, B.D., Kriesky, J., & Pavliakova, B. 2012. Missing from the table: Role of the environmental public health community in governmental advisory commissions related to Marcellus Shale drilling. *Environmental Health Perspectives*, 120(4), 483–486.
Götschi, T., Heinrich, J., Sunyer, J., & Künzli, N. 2008. Long-term effects of ambient air pollution on lung function: A review. *Epidemiology*, 19(5), 690–701.
Gould, S. 1981. *The Mismeasure of Man*. New York: Norton.
Hertz-Piccioto, I. 1998. 'Environmental Epidemiology' in Rothman, K.J., & Greenland, S. (eds.). *Modern Epidemiology*. (2nd ed.) Philadelphia: Lippincott-Raven Publishers. pp 555–583.
Hill, A.B. 1965. The Environment and disease: Association or causation? *Proceedings of the Royal Society of Medicine*, 58(5), 295–300.
Kasperson, R., Renn, O., Slovic, P., et al. 1988. The social amplification of risk: A conceptual framework. *Risk Analysis*, 8, 177–187.
Kimmel, C. 1990. Quantitative approaches to human risk assessment for non-cancer health effects. *Neurotoxicology*, 11(2), 189–198.
Lagakos, S., Wessen, B., & Zelen, M. 1986. An analysis of contaminated well water and health effects in Woburn, Massachusetts. *Journal of the American Statistical Association*, 81, 583–596.

Lake, I.R., Nichols, G., Bentham, G., Harrison, F.C.D., Hunter, P.R., & Kovats, R.S. 2007. Cryptosporidiosis decline after regulation, England and Wales, 1989–2005. *Emerging Infectious Diseases*, 13(4), 623–625.

Lambert, T.W., Guyn, L. & Lane, S.E. 2006. Development of local knowledge of environmental contamination in Sydney, Nova Scotia. *Science of the Total Environment*, 368(2–3), 471–484.

Leedy, P., & Ormrod, J. 2001. *Practical Research: Planning and Design*. (7th ed.) Upper Saddle River, NJ & Thousand Oaks: Merrill Prentice Hall and SAGE Publications.

Levin, K. 2006. Study design III: Cross-sectional studies. *Evidence-Based Dentistry*, 7, 4–5.

Medical Research Council (MRC). 2009. *Using Natural Experiments to Evaluate Population Health Interventions*. Available at: http://www.mrc.ac.uk/documents/pdf/natural-experiments-guidance/ (Accessed 3 June 2015).

Moeller, D.W. 2009. *Environmental Health*. Cambridge: Harvard University Press.

Morse, J. 1995. The significance of saturation. *Qualitative Health Research*, 5, 147–149.

National Research Council (NRC). 2006. *Toxicology Tests for Environmental Agents*. Washington, DC: National Academies Press.

Newland, M.C., & Paletz, E.M. 2000. Animal studies of methylmercury and PCBs: What do they tell us about expected effects in humans? *Neurotoxicology*, 21(6), 1003–1027.

Ochodo, C., Ndetei, D.M., Moturi, W.N., & Otieno, J.O. 2014. External built residential environment characteristics that affect mental health of adult. *Journal of Urban Health*, 91(5), 908–927.

Oreskes, N. 1998. Evaluation (Not validation) of quantitative models. *Environmental Health Perspectives*, 106(Suppl 6), 1453–1460.

Pruss, A. 1998. Review of epidemiological studies on health effects from exposure to recreational water. *International Journal of Epidemiology*, 27(1), 1–9.

Reedijk, M., Van Leuken, J.P.G., & Van der Hoek, W. 2013. Particulate matter strongly associated with human Q fever in the Netherlands: An ecological study. *Epidemiology and Infection*, 141(12), 2623–2633.

Rothman, K.J., & Greenland, S. 2005. Causation and causal inference in epidemiology. *American Journal of Public Health*, 95, S144–S150.

Sabatier, P.A. 1987. Knowledge, policy-oriented learning, and policy change an advocacy coalition framework. *Science Communication*, 8(4), 649–692.

Schmidt, C.O., & Kohlmann, T. 2008. When to use the odds ratio or the relative risk? *International Journal of Public Health*, 53(3), 165–167.

Shepherd, D., McBride, D., Dirks, K.N., & Welch, D. 2014. Annoyance and health-related quality of life: A cross-sectional study involving two noise sources. *Journal of Environmental Protection*, 5, 400–407. doi: 10.4236/jep.2014.55043

Sugiri, D., Ranft, U., Schikowski, T., & Krämer, U. 2006. The influence of large-scale airborne particle decline and traffic-related exposure on children's lung function. *Environmental Health Perspectives*, 114(2), 282–288.

Thayer, K.A., Heindel, J.J., Bucher, J.R., & Gallo, M.A. 2012. Role of environmental chemicals in diabetes and obesity: A national toxicology program workshop review. *Environmental Health Perspectives*, 120, 779–789.

Tola, S., Vilhunen, R., Järvinen, E., & Korkala, M.L. 1980. A cohort study on workers exposed to trichloroethylene. *Journal of Occupational Medicine*, 22(11), 737–740.

Venables, D., Pidgeon, N., Simmons, P., Henwood, K., & Parkhill, K. 2009. Living with nuclear power: A Q-method study of local community perceptions. *Risk Analysis*, 29, 1089–1104.

Wartenberg, D., Reyner, D., & Siegel Scott, C. 2000. Trichloroethylene and cancer: Epidemiologic evidence. *Environmental Health Perspectives*, 108(Suppl 2), 161–176.

Williams, C. 2007. Research methods. *Journal of Business & Economic Research*, 5(3), 65–72.

Williams, D.R., Neighbors, H.W., & Jackson, J.S. 2003. Racial/Ethnic discrimination and health: Findings from community studies. *American Journal of Public Health*, 93(2), 200–208.

Zilsel, E. 1942. The sociological roots of science. *American Journal of Sociology*, 47, 544–562.

2 Uncertainty, social science and the role of theory

Introduction: Risk, uncertainty and the role of theory

"Risk is to do with uncertainties: possibilities, chances, or likelihoods of events, often as consequences of some activity or policy. As such, risk has always accompanied the development of human society" (Taylor-Gooby and Zinn, 2006, p. 1). Risk and uncertainty increasingly involve social and political frameworks and institutions to understand and manage them – often in terms of what is acceptable or not, leading to conflict and contestation. Some of these institutions themselves appear threatening, and there has been a significant questioning of expertise in mapping out risks, the role of uncertainty and what is regarded as safe and beneficial. We will discuss some of these issues in Chapters 3 and 4. Contesting the role of government and the increasing sophistication of probability assessments have gone hand in hand, making any assessment of risk and uncertainty both a technical and a political matter. There remains significant ambiguity in terms of the meaning of risk. A neutral term from the past to evaluate gains and losses from particular actions has now become weighted towards loss. "Risk is the probability that a particular adverse event occurs during a stated period of time, or results from a particular challenge" (Royal Society, 1992, p. 4). Risk involves chance and safeguards against risky events or activities. In fact, safety is seen as freedom from unacceptable risks of personal harm. And as Hayes (1992) argues in a polemic, social science approaches alter definitions, characteristics and responses to risk. Language is everything and its use

> by different groups to describe risk and to prescribe solutions is judgmental. The terms employed frame an event; the metaphors and images used to describe a situation can point the finger of blame and imply responsibility for remedial action. Some words imply disorder or chaos, others certainty and scientific precision. Selective use of labels can trivialize an event or render it important; marginalize some groups, empower others; define an issue as a problem or reduce it to a routine.
>
> (Nelkin, 1985, cited by Hayes, 1992, p. 403)

Nelkin (1985) speaks of environmental hazards being seen as a fact of life or something to produce panic or phobias. Little has changed in the 30 years since her writing. These panics can now be seen as central to understanding risk, uncertainty and social responses.

Risk and panics

Panics are seen by many as 'irrational' responses to events (Taubes, 1995). Ungar (2001) notes how panics and social anxieties in advanced industrial societies have built up around nuclear, chemical, environmental, biological and medical issues. These anxieties are fuelled by the existence of environmental or technological possibilities for disaster so that risk aversion and the desire to find safe haven or scapegoats are reinforced. This is Beck's risk society in action (Beck, 1992) in which a situation is amplified so that hazards and their attributes (e.g., deaths, injuries, damage, and social disruption) are real enough, these interact with a wide range of psychological, social, or cultural processes in ways that transform signals about risk (see Kasperson and Kasperson, 2005, and Chapter 5). These signals are filtered by scientists, the media (conventional and social), politicians, government agencies and other interest groups to intensify or amplify the risk. In risk society, the discourse of safety becomes problematic. This

> discourse faces rupture in the risk society. Invisible contaminants, intractable scientific uncertainties, unpredictable system effects, the almost tragic calls for 'science-on-demand' at the height of an accident, the prying open of standard operating procedures, efforts to pass off the hot potato, and potential latency effects that hinder closure of the threat.
>
> (Ungar, 2001, p. 287)

One example in which all these elements appear is the case of contaminated drinking water in Walkerton, Ontario, Canada, in 2000. In a farming town of fewer than 5,000 people, 160 sought hospital treatment and another 500 phoned the hospital complaining of vomiting, cramps and diarrhoea. This 24- to 36-hour event was soon overtaken by a larger tragedy of eventually 2,300 becoming ill and 7 people dying from what was diagnosed as the first *E. coli O157:H7* infection of a town water supply. A judicial inquiry was established, and a journalist wrote a book on the situation (Attorney General, 2002; Perkel, 2002). The agricultural economy with its factory farming of livestock with enormous animal waste production was implicated. So too were the local operators of the drinking water system, who became the scapegoats with less attention being paid to the privatization of water supply management or of water testing and monitoring labs with little government oversight, this being a time of strenuous rolling back of the state. Furthermore, it took time to isolate the causes of the gastrointestinal problems and other health effects suffered by the local population. Some science is not rapid. While the sicknesses in 2000 could be strongly associated with the polluted water, long-term effects took 5 to 10 years and more to determine and with mixed evidence (see Garg et al., 2005; Villani et al., 2010). What received less press coverage is that the social structure of the town itself was also subtly implicated in a slower response than was needed due to pride in local water sovereignty (Parr, 2005).

The Walkerton disaster did lead to significant policy change with respect to water quality monitoring. Underlying the failures of the Walkerton public utility and the environment ministry were the government of Ontario cutbacks. How deep were the cuts? In the years leading up to the Walkerton tragedy, the Ministry of Environment's (MOE's) budget was reduced by 68% and its staffing by 40%. A Clean Water Act was passed after the disaster and enquiry, but local response has been slow. With respect to compensation, CTV (2011) indicated some 10,189 claims were made, with 9,275 qualifying. By mid-2011, Ontario had paid out $72 million to the victims and their families. With respect to health, those who came down with acute gastroenteritis from the tainted water were more likely to develop hypertension or kidney problems or have a cardiovascular event, compared with those who were not ill or only mildly ill. Science moves slowly.

Assessing or estimating risk – probabilities – statistical theory – more science in better estimates, application of influence of genes

Despite the panics, science remains firmly embedded as a key response to assessing risk. Indeed, in an era of invisible and dreaded risks, it is paradoxically paramount. Furedi (1997) expresses this well: an enhanced sense of risk consciousness means that concerns of caution, fear and danger are not so much about what is morally acceptable or proper but become discourses about safety and control. So scientific assessment is both what is not needed and what is needed.

Pidgeon et al. (2006, p. 95) sum up the challenges of a scientific approach to risk assessment. Environmental and technological issues tend to have a number of characteristics often difficult to deal with both in formal risk assessments and in public policy. The authors add that many environmental threats stem from very low-probability but high-consequence events (Pidgeon et al., 2006). Not all environmental risks can be described probabilistically, and there is uncertainty including at the expert level. Environmental and technological risk invariably presents both hazard and opportunity, and it is now well known that people evaluate outcomes very differently depending upon whether they personally view them as 'losses' or 'gains'. Many hazards, such as naturally occurring radon gas, will never have been directly experienced by those who must be persuaded to act to mitigate their very real future consequences. And environmental hazards often also involve making difficult trade-offs over time, with consequences possibly very far into the future, such that long-term effects are inequitable in their distribution across different groups or must be anticipated for people not yet born (e.g., burning oil and coal offers benefits now but brings future risk from climate change). Scientific assessment tries therefore to become more rigorous, all encompassing and more detailed. This may be noted in formal risk assessments where uncertainty is managed by the use of probability theory.

In a recent, often seen as path-breaking book on probability, Jaynes (2003) notes that the goal of inference in this risk assessment should be to estimate not only the slope of the response curve but, far more importantly, to decide whether there is evidence for a threshold and, if so, to estimate its magnitude (the 'maximum safe dose'). Calculations based on adequate sampling, prior occurrences and deductive reasoning lead to plausible conclusions with uncertainties reduced to specific probabilities allowing statements about safety and risk to be stated with certain degrees of confidence. We put on one side how it is next to impossible to avoid value and emotion entering into those events and activities we consider worthy of investigation. So what is the likelihood risk will occur? For environmental health risk, we are also concerned about possible impact – the level of dose. Importance of dose or impact, severity and consequences allows for consideration of vulnerability, equity and the non-human world, all of which require judgments. But as we shall see there are scientific and economic appraisals of these concepts. We explore these ideas further in Chapter 9.

Yet we note that in carrying out these assessments difficult decisions have to be made on exposure source, route-ways, safe dose and so on, often on a community or aggregate level, even if the outcome affects specific types of individuals. Such assessments tend to be quite conservative in applying the logic of scientific investigation. For example, a complex study of the likely impact of BTEX from landfill sites concluded that emissions do not pose a health threat to workers at the site. Furthermore air dilution of BTEX emitted from the site is sufficient to guarantee the protection of population in the local neighbourhood (see Durmusoglu et al., 2010). This conservatism makes positive associations difficult to find, and as we shall see, scientists prefer false negatives to false positives. It is perhaps better to measure more phenomena to try to tease out relationships, which may make assessment and policy development more difficult. There has, for example, been much recent work on gene–environment interactions and their possible health impact. Many diseases result from the interaction of microorganisms and individual genetic make-up. For example, use of two pesticides, rotenone and paraquat, were linked with Parkinson's disease. People who often used either pesticide developed Parkinson's disease approximately 2.5 times more often than non-users in Canada and more stringent bans in China and the EU, in places where its use is still allowed (see Tanner et al., 2011). Further, as we indicate further in the chapter, occupational exposure – for example, where a substance is manufactured or applied – for many exposures is a good first step in assessing health impacts, since those exposures tend to be highest for human populations.

Assessing environmental health risk in risk management

Scientific assessments remain an important dimension in risk management, although there have been a few decades of recognition that assessment and risk estimation are one part of understanding and managing risk (see Figure 2.1). In

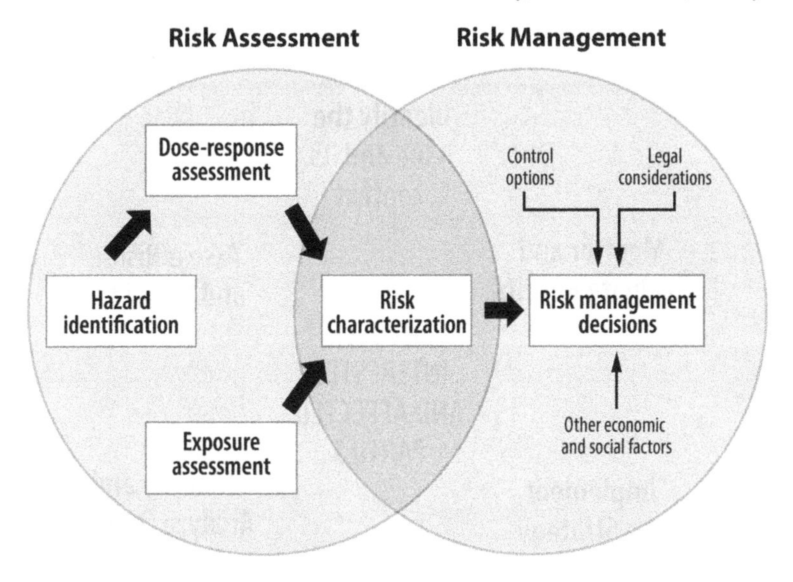

Figure 2.1 Connecting risk assessment and risk management

Source: Adapted from EPA (2014).

the United States, for example, risk assessment has been in use since the 1950s but has become more sophisticated and accurate with increased scientific rigour and increasing interest from government regulators. In the 1960s and 1970s, the Environmental Protection Agency (EPA) and the Occupational Safety and Health Administration (OSHA) were formed under U.S. federal legislation, along with adoption of numerous laws regulating environmental hazards. But resources had to be allocated among many competing issues, so they needed tools to help them focus on the most dangerous hazards. Former EPA administrator William K. Reilly recalled,

> Within the space of a few years, we went to the possibility of detecting not just parts per million but parts per billion and even, in some areas, parts per quadrillion . . . That forces you to acknowledge that what you need is some reasonable method for predicting levels of real impact on humans so that you can protect people to an adequate standard.
>
> (EPA, no date)

In 1983, the U.S. government developed a way forward (NRC, 1983). This was later extended in 1996 in an NRC document on understanding risk, adopted by other jurisdictions too, including Canada, to include not only assessment but options, communications and evaluation (PHAC, 2013; Figure 2.2).

Briggs (2008) argues that in assessing health problems linked to the environment much depends on “how issues are selected and framed by all stakeholders

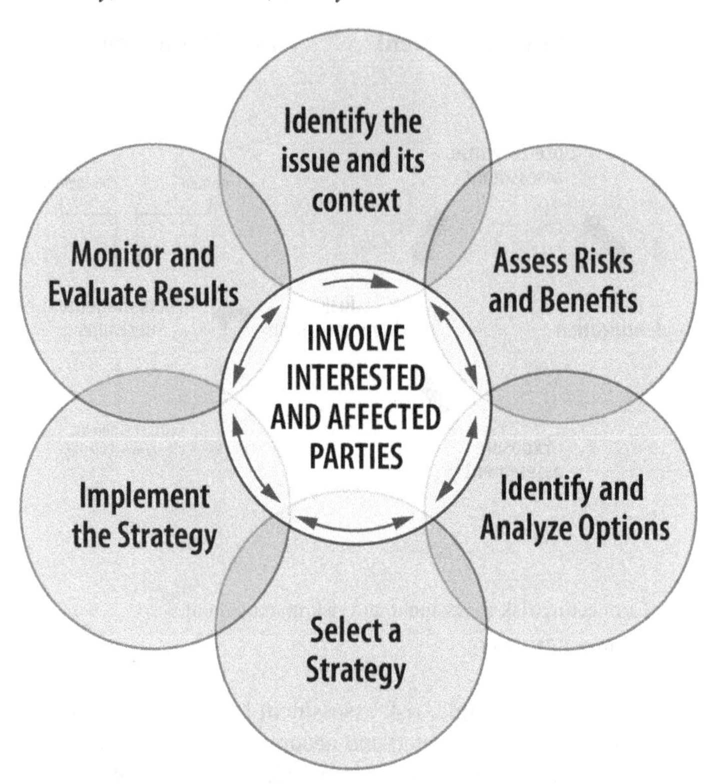

Figure 2.2 Public Health Agency of Canada framework for understanding risk and public policy (PHAC, 2006)

Source: Strategic Risk Communications Framework within the Context of Health Canada and the PHAC's Integrated Risk Management. Public Health Agency of Canada, 2006. Reproduced with permission from the Minister of Health, 2015.

and involving evaluations of options" (see Figure 2.3). He makes key points about differences in what integrated risk assessment is compared with health impact assessment, with the former being used to assess the risks from multiple exposures associated with, say, endocrine-disrupting agents, by considering agents that have a common mechanism or effect. A similar approach has been identified for persistent organic pollutants (see also Ross and Birnbaum, 2003; Bridges and Bridges, 2004). Health impact assessment focuses on policies, facilities and interventions rather than on agents. It also recognises that the environment may be a hazard, but it may provide benefit by providing natural capital or ecological services, for example water security, green spaces (see Mindell and Joffe, 2003). We would add then that societal attitudes toward 'nature' become extremely important in HIA as part of issue identification (see Chapter 9). Briggs (2008) goes on to develop an integrated EHIA which takes into account such issues and how they help shape risk

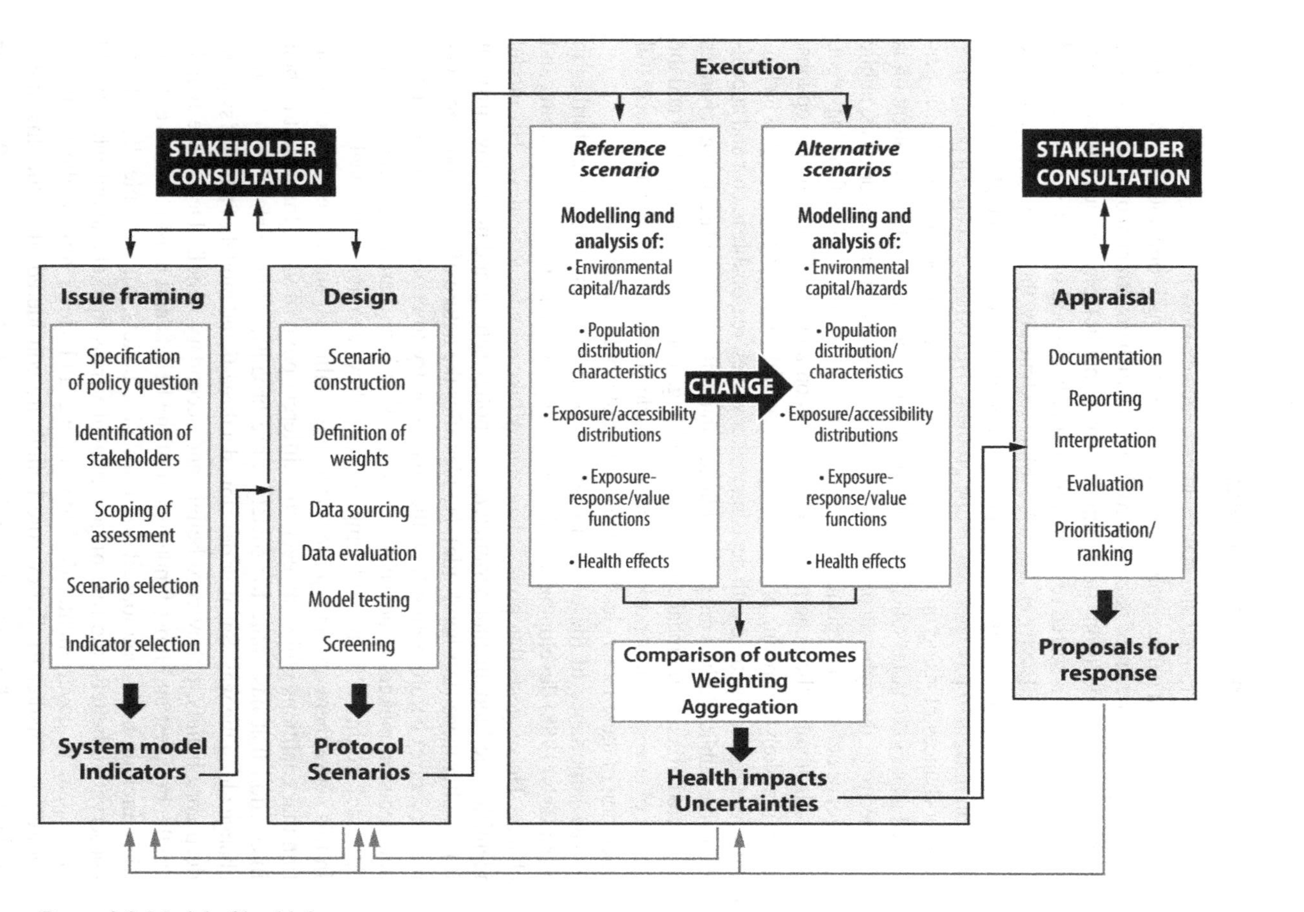

Figure 2.3 Model of health impact assessment

Source: Briggs 2008 licensed under CC BY 4.0 from BMC Environmental Health.

governance (Figure 2.3). For EHIAs the left-hand box – Issue Framing – is vital since it sets the stage for what is to be considered and what is not in the assessment.

Issue framing and agenda setting

So how are issues framed and stakeholders selected? How is the agenda set? Kingdon (1995) sets out a three-stage approach. Agenda setting is only the first stage. The policy agenda is the list of issues or problems to which government officials, or those who make policy decisions (including the voting public), pay serious attention. Yet public attention is a scarce resource unless life, livelihood or values are threatened (Eyles et al., 1993; Baxter et al., 1999). More technically, three processes can come into play: problems, proposals and politics. Problems refer to attempts to persuade decision makers to pay attention to one problem over others. Because a policy proposal's chances of rising on the agenda are better if the associated problem is perceived as serious, issue salience is critical. It can be influenced by how attention is obtained (e.g., through data or indicators, focusing events – a disaster or crisis, public outrage) or defined. Threats to life from environmental hazards are a focusing event. Proposals involve the generation, debate, revision and adoption of policy options. There are likely competing proposals, and successful ones are usually seen as technically feasible, compatible with decision maker values, reasonable in cost and appealing to the public. Politics are political factors that influence agendas, such as changes in elected officials, political climate or mood (e.g., conservative, tax averse) and the voices of advocacy or opposition groups. Kingdon (1995) argues that the successful operation of these factors leads to the opening of a policy window.

The importance of these dimensions for policy success was taken further in Sabatier's (1988) development of the 'advocacy coalition' for policy learning and change. He suggests that policy change may take decades to achieve. His model identifies the fundamental parameters of a policy setting and notes how difficult it is to alter these (left-hand side). He identifies the nature of operational engagement (centre) before noting how the coalitions try to engage others for support and decision makers to act in their favour. This is an iterative process, and the advocacy coalition framework emphasizes policy learning as an ideal outcome. In this sense there are no winning or losing coalitions, and there is no expectation that coalitions will sway each to a different point of view. Instead, it is their interaction that has value for society more broadly, while surface beliefs may change along the way. Sabatier developed a three-tiered model of a belief system: deep core beliefs, policy core beliefs and secondary beliefs. Deep core beliefs are the broadest and most fundamental of the beliefs and most stable over time. They include basic ontological and normative beliefs, for example, liberal and conservative beliefs, the role of government versus markets, about who should participate in governmental decision making and views toward the resiliency nature (see Chapter 9). Policy core beliefs are considered 'the glue' that holds coalitions together as they represent basic normative commitments to, and perceptions of, problem definition, the causal mechanisms and the appropriateness of institutional arrangements to deal with the policy issue. Weber et al. (2013)

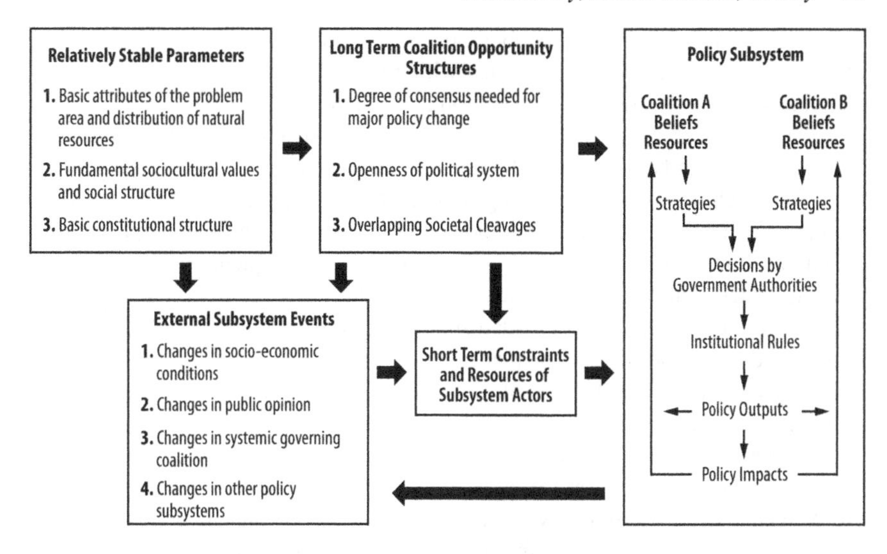

Figure 2.4 Sabatier's policy and advocacy coalition framework

Source: Sabatier (1988), with kind permission from Springer Science and Business Media.

successfully used the advocacy coalition framework (ACF) to analyze noise policy in the Netherlands, identifying three coalitions – industry, environmental health and planning – with a lack of trust paralyzing the development of noise policy. ACF has provided a useful framework in pesticide policy change in Calgary and Halifax, with certain coalitions having greater access to decision makers than others (see Hirsch et al., 2010).

Values in assessing risk

From these ideas, the importance of values and mind-set emerges. In some cases, there is conflict about how individuals and groups regard environmental hazards. A common one concerns the relative importance of environmental degradation and health on the one hand and economic factors (jobs, investment, trade agreements) on the other. So how is environment/nature treated? Marx explored the environment through the lens of the American experience to comment on the true natural and industrial potential of a "virgin continent" (1963, p. 3). He tried to resolve human social and psychological response to changing technology into a "grand design" involving the interplay between two contrasting influences – the "pastoral" and the "counterforce". The pastoral represents not only a country or garden scene, as in its contemporary usage, but also, more generally, any single ideal conception of peace or harmony put forth in painting, music, literature or simply as one of many "certain products of the collective imagination" (Marx, 1963, p. 4). In other words, it is nature as it is imagined – a key element in environmentalist thinking. The 'counterforce', "whether represented by the plight of a dispossessed herdsman or by the sound of a locomotive in the woods . . . brings a world which is more 'real' into juxtaposition with an idyllic vision" (Marx, 1963, p. 25). This is the machine in the garden which provides great benefits for most humankind in terms of their well-being, longevity and physical health.

This idea of a machine age is taken up by Lynch (1984), who sets out three theories used to understand environmental–human interfaces, that is religious (the grand design), the machine (human will being dominant) and organic (environment as a living thing). How the environment is viewed shapes strongly how its impacts are regarded and how it is treated whatever the long-term consequences. These ideas can be illustrated with examples which benefited the health of populations (e.g., Zuider Zee in the Netherlands) or traded short-term economic and social gains for long-term and lasting adverse consequences (e.g., the Aral Sea in central Asia).

Formerly one of the four largest lakes in the world with an area of 68,000 km^2 (26,300 sq miles), the Aral Sea has been steadily shrinking since the 1960s after the rivers that fed it were diverted by Soviet irrigation projects (see Figure 2.5). By 2007, it had declined to 10% of its original size, splitting into four lakes – the North Aral Sea, the eastern and western basins of the once far larger South Aral Sea and one smaller lake between the North and South Aral Seas.

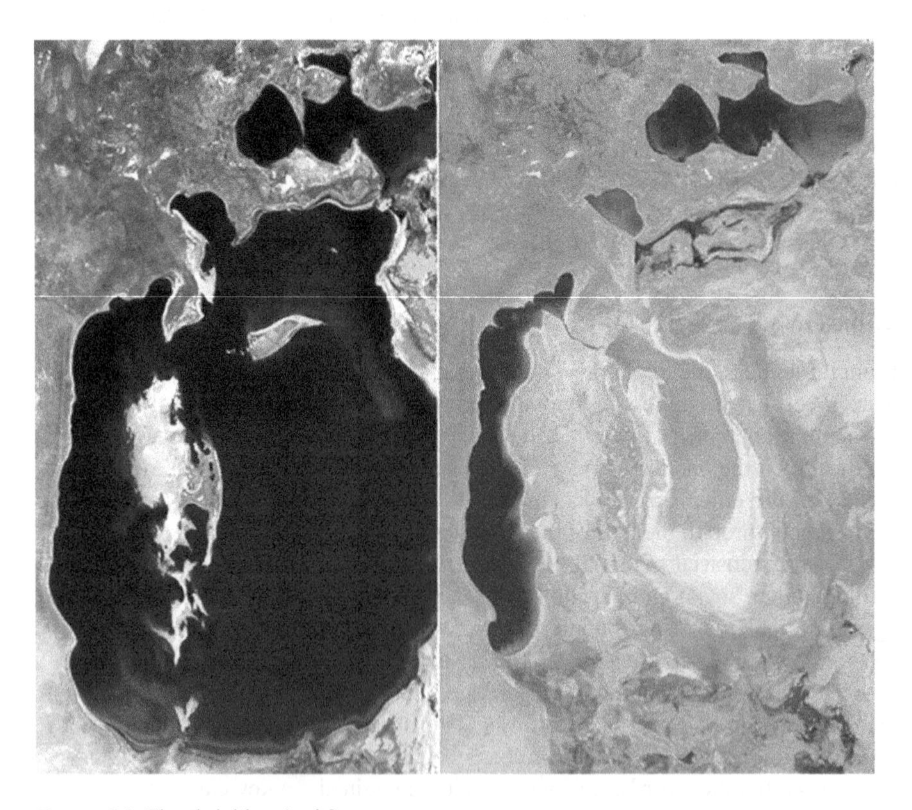

Figure 2.5 The shrinking Aral Sea

Source: NASA (2014). The Aral Sea, Earth Observatory https://img.rt.com/files/news/2e/ca/80/00/nasa.jpg

The ecosystems of the Aral Sea and the river deltas feeding into it have been nearly destroyed, with much higher levels of salinity. The receding sea has left plains covered with salt and toxic chemicals – the results of weapons testing, industrial projects and pesticides and fertilizer runoff – which become airborne. There is a lack of fresh water with significant metal contamination (see Törnqvist et al., 2011). There are many health problems, including high rates of certain forms of cancer and lung diseases. Respiratory illnesses, including drug-resistant tuberculosis, digestive disorders, anaemia and infectious diseases are common. Liver, kidney and eye problems can also be attributed to the toxic dust storms. Health concerns associated with the region are a cause for an unusually high fatality rate amongst vulnerable parts of the population. The child mortality rate is 75 in every 1,000 new-borns, and maternity death is 12 in every 1,000 women (see for example, Whish-Wilson, 2002).

On a smaller scale, 'Cancer Alley' in Louisiana between Baton Rouge and New Orleans can be cited. In a case of popular epidemiology, citizens noted a large number of cancer deaths in close proximity to one another. The area has around 140 petro-chemical plants, operating in poor, largely black parishes (counties). These plants form a major part of the tax base of Louisiana, and new permissions have been granted to build more. St. James Parish, close to New Orleans, is considered by the Environmental Defense Fund to be one of America's 25 most polluted counties, Louisiana had the second-highest death rate from cancer in the U.S. in 2002; it had also been ranked second in the U.S. by the EPA for total on-site releases of chemicals and pollutants. In some ways, this may be seen as the revenge of the machine, with industrial activity creating jobs and potentially ruining health. The matter is contentious, whereby, for example, white males in the area are less apt to perceive local industry represents a serious health threat, while black women are most likely to do so (Brent, 2006). One of the companies funded research which showed cancer rates in the area are lower than state and national levels (see Tsai et al., 2004). Other research (e.g., Billings, 2005) has shown that tobacco use confounds the relationship between environmental hazard and health. Yet cancer outcomes may be beside the point, as there is compelling evidence that black communities are disproportionately exposed to facilities compared to white counterparts in the local area, particularly white people who work in the petrochemical industry (Blodgett, 2006).

These examples begin to frame issue identification – the first dimension in any risk assessment – as involving power, interests and equity matters. Thus vulnerability has become a key political economic concept in assessing risk. As Füssel (2007) notes, vulnerability is difficult to define and requires consideration of a population and an event or situation. The United Nations (2004) distinguishes four vulnerability factors: physical – the exposure of vulnerable elements within a region; economic – the economic resources of individuals, population groups and communities; social – non-economic factors that determine the well-being of individuals, population groups and communities, such as the level of education, security, access to basic human rights and good governance; and environmental – the state of the environment within a region. These collectively create the conditions for vulnerability. Yet as Morrow (1999) notes, disaster vulnerability is socially constructed, that is, it arises out of the social and economic circumstances of everyday living. Her introduction of constructionism is important (see what follows), as it focuses more on the experiences with a given system. Turner et al. (2003) note that vulnerability lies in the interaction between social dynamics and their containing social-ecological system. It is not necessarily a disaster situation. In a constructivist mode, those who perceive themselves to be vulnerable to environmental risks or who see themselves to be victims of injustice also feel more at risk from environmental hazards (Satterfield et al., 2004). This is further complicated by the notion that perceptions of barriers to adapting by the vulnerable do in fact limit adaptive actions, even when there are capacities and resources available to adapt (Grothmann and Patt, 2005). Those with little hope, who see themselves as downtrodden and at the bottom of the social pile, may lack resilience and get taken by surprise by changes in hazard exposure.

Many studies have documented the disproportionate location of hazardous waste sites, industrial facilities, sewage treatment plants and other locally undesirable and potentially polluting land uses in communities of racial or ethnic minorities and in socially disadvantaged neighbourhoods (see Mohai et al., 2009). As Morello-Frosch et al. (2011, p. 881) comment,

> The residents of communities near industrial and hazardous waste sites experience an increased risk of adverse perinatal outcomes, respiratory and heart diseases, psychosocial stress, and mental health impacts. Members of racial or ethnic minority groups and people of low socioeconomic status are also more likely than others to live near busy roads, where traffic-related air pollutants concentrate.

Morello-Frosch and Shenassa (2006) have identified a cumulative environmental riskscape for racial minorities and lower classes that add to the vulnerability burden.

> The cumulative physiological 'wear and tear' resulting from chronic overactivity of the body's stress-response system may impair immune functioning and increase vulnerability to stressors by increasing the absorption of toxicants into the body through increased respiration, perspiration, and consumption; compromising the body's defense systems against toxicants; affecting the same physiological processes as environmental agents; and directly causing illness.
>
> (Morello-Frosch et al., 2011, p. 882)

Burden of proof for vulnerability and health damage remains largely with the community, although local initiatives have developed around food security, access to green spaces and so on. Cutter (1996) notes the discourse of hazard management, dominated by technical and engineering approaches, failed to engage with the political and structural causes of vulnerability. Adger (2006) cites Hewitt's idea of the human ecology of endangerment, allowing him to be critical of the pressure-release models in which socio-economic conditions and physical exposure can result in vulnerability. Yet these models do link vulnerability and risk well and permit an understanding of likely surprise and resilience. There is then a progression to vulnerability (see Blaikie et al., 1994; Wisner et al., 2004; Figure 2.6).

Vulnerabilities arise not only from 'natural' hazards but can be human produced or induced, that is human activities which worsen natural disasters. For example, New Orleans was built on a natural high ground along the Mississippi River. Later developments, extended to nearby Lake Pontchartrain, were constructed on fill to bring them above the average lake level. Navigable commercial waterways extended from the lake into the interior of the city to promote waterborne commerce. After the construction of the Inner Harbor Navigation Canal in 1940, the state closed these waterways, causing the town's water table to lower drastically. After 1965, the United States Army Corps of Engineers built a levee system around a much larger geographic footprint that included previous marshland

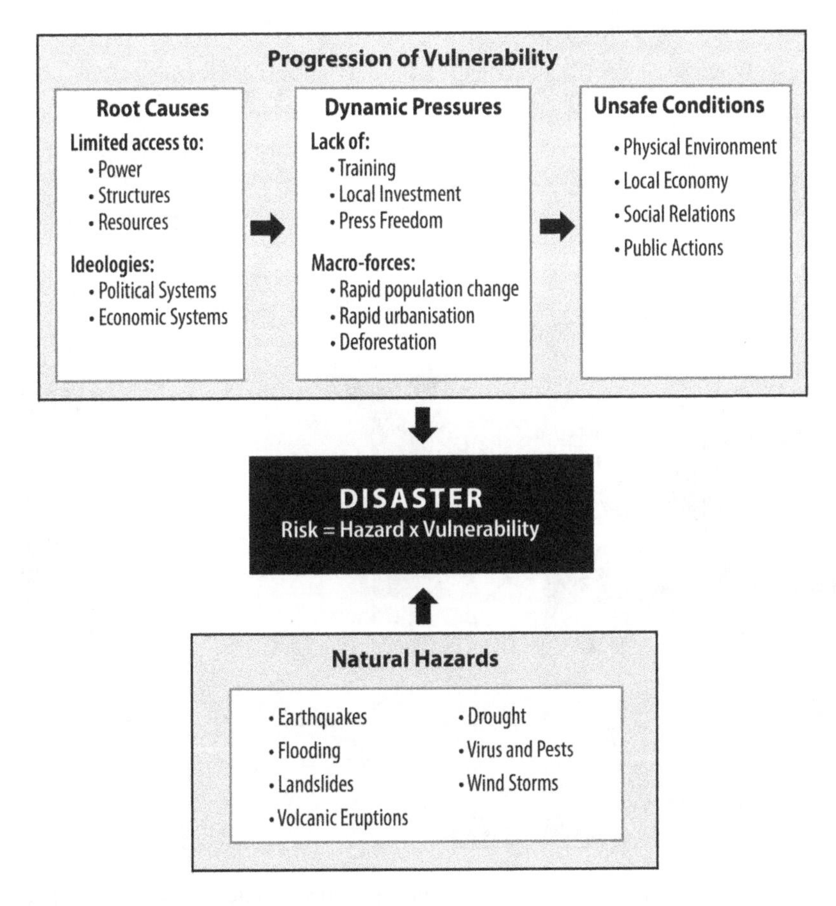

Figure 2.6 Social vulnerability framework

Source: Wikimedia Commons after Blaikie et al. (1994).

and swamp. In 2005, the storm surge from Hurricane Katrina caused more than 50 breaches in drainage and navigational canal levees and precipitated the worst engineering disaster in the history of the United States (see Figure 2.7).

The flooding created not only economic disaster but local hospital closures and an increase in infectious diseases and some deaths (Vince, 2005). Mismanagement and poor leadership in Federal Emergency Management Agency (FEMA) as

Figure 2.7 Flooding after Hurricane Katrina in New Orleans: Interstate 10 at West End Boulevard, looking towards Lake Pontchartrain

Source: U.S. Coast Guard (2005).

well as local and state governments is considered responsible, while reconstruction is slow (see Biever, 2009) and partial. Not all can still be protected by new canal levees and dikes. The tourist quarter and wealthy suburbs seem to have more than their fair share. As *The Guardian* ("New Orleans," 2014) reported,

> In the wake of the flood, money surged into New Orleans. FEMA provided nearly $20bn to help Louisiana recover from Katrina and Rita, the hurricane which followed a month later. The state's Road Home program claims to have channeled about $9bn in government funds to more than 130,000 residents. But Road Home was beset by administrative problems, and the subject of a discrimination lawsuit settled in 2011 because, typically, it handed out grant money based on a home's pre-Katrina worth, rather than the cost of fixing the damage, advantageous to people who lived in more expensive, white areas.

Parts of the poor, minority Ninth Ward remain barren and largely untouched. Why is this the case? Why does maladministration affect the poor more?

Perhaps it is not so much that the poor are deliberately targeted, it is more that those with power continue to receive benefit and are less affected by harm. It is not just their economic interests but their ability to define and shape what is harmful and the interventions to deal with harm and costs – all of which are shaped by how they themselves live and work. These inequities have been identified and discussed by academic researchers and environmental justice groups (See Chapter 8). Since Bullard's (1990) study of 'dumping in Dixie' many studies have identified inequities in environmental risk and in health problems, a situation often supported by industrialists and developers claiming to create jobs (see Been, 1992; Schlosberg, 2013). Beneath these inequities and appeals for justice lurk interests and a dominant ideology, often supported by specific institutions. These political ideas have been mentioned before in relation to the ACF and values in risk assessment. But institutions are themselves constituted by interests and beliefs or ideologies and form in a sense cognitive models for assessing and managing risk. Emerging economies perhaps illustrate best of all this linking on vested interests, ideologies for economic development (often at the expense of environment and health) and the importance of institutions to further these (see Xu, 2011, on the Chinese experience). Oreskes and Conway (2010) provide examples of how vested interests attempt to shape the scientific assessments of such risks as second-hand smoke, acid rain and the use of some pesticides. In such cases, as we shall exemplify in later chapters, the demand for evidence-based assessments becomes a challenge to risk assessment and management themselves. In their conclusion, Oreskes and Conway (2010, p. 274) cite a past director of research for British American Tobacco: "a demand for scientific proof is always a formula for inaction and delay, and usually the first reaction of the guilty. The proper basis for . . . decisions is . . . that which is reasonable in the circumstances." Thus, they suggest more, for example, epidemiologic research is not always the answer to environment and health policy dilemmas, and a hard look at who is negatively impacted by current social arrangements may be sufficient to warrant action.

Recognizing public values

Analysts often argue that risk assessments are technical, not dealing with values. But all assessments are normative, even when there is widespread scientific and public agreement. All scientific investigation is unavoidably laden with theory because it requires both a definition of the research problem and ways of assessing the rigour of relevant evidence that relies on particular normative standards (Hill's criteria are often used). In the case of GMOs, traditional risk assessments of agricultural biotechnologies regard GMOs as decontextualized biological entities that can be subject to empirical investigations (see de Melo-Martin and Meghani, 2008). Although, of course, this is perfectly legitimate under particular contexts, such assessments are problematic because they overlook the fact that those seeds are also objects of social value with economic, legal, cultural or aesthetic significance (Lacey, 2005).

Risk assessments of GMOs also raise ethical questions: what counts as a serious risk? What is the relevant time frame for investigating such risks? What are the standards required to judge that unmanageable risks are not present? What is an acceptable level of risk? Moreover, social affiliations – such as profession, gender and political ideologies – influence what determines a risk (Douglas and Wildavsky, 1982). Not surprisingly, the views of laypeople and experts as to what constitutes a risk are often different (Savadori et al., 2004). Laypersons tend to value both the context of risk as well as its content, whereas experts usually place greater emphasis on risk endpoints rather than their context (see later chapters). So it is important to understand how the public evaluates risk and hazard exposure.

The next two chapters comment on the public evaluation of exposures. In fact all our chapters are themed by the German Advisory Council on Global Change (2000) scheme to evaluate risks. It outlined the following dimensions:

- Estimate of event/damage and probability of occurrence – estimate/assignment of the relative frequency of an event
- Incertitude – overall indicator for different uncertainty components
- Ubiquity – defines the geographic dispersion of potential damages (intragenerational justice)
- Persistence – defines the temporal extension of potential damage (intergenerational justice)
- Reversibility – describes the possibility of restoring the situation to the state before the damage occurred (possible restoration – e.g., reforestation and cleaning of water)
- Delay effect – characterizes a long time of latency between the initial event and the actual impact of damage; the time of latency could be of physical, chemical or biological nature
- Violation of equity – describes the discrepancy between those who enjoy the benefits and those who bear the risks
- Potential of mobilization – is understood as a violation of individual, social or cultural interests and values, generating social conflicts and psychological

reactions by individuals or groups who feel affected by the risk consequences; they could also result from perceived inequities in the distribution of risks and benefits.

These risk evaluations categorize public values which may increase or decrease uncertainty, fear and panic about hazards. In sum, they are shaped by our overarching theory – social constructionism, which can itself be given differentiation by social evaluation theory – the importance of comparing ourselves with others.

Overarching theories: Social constructionism and social evaluation

Social evaluation – the way that people learn about themselves by comparing themselves with others – is an age-old process. Gartrell (1987) demonstrates that social evaluation involves reference groups, social comparison, equity and justice and relative deprivation. It is thus a key dimension of understanding how individuals and communities assess hazard and risk. As we have already noted, social evaluations are types of judgment in which what is hazardous or not is determined. Judgment is shaped by information and the legitimacy of the organizations providing information or interventions (see Bitektine, 2011). This legitimacy can shape the kinds of social evaluations and judgments made as well as the reinforcement of belief systems and past response mechanisms to hazards. Later chapters will deal with legitimacy issues and their relationship to risk characterization and response. But all these dimensions may be subsumed as the social construction of risk.

Social constructionism tries to uncover the ways in which individuals and groups participate in the construction of their perceived social reality. It involves looking at the ways social phenomena are created, institutionalized, known and made into traditions and/or myths. The social construction of reality is an ongoing, dynamic process that is reproduced by people acting on their interpretations and their knowledge of it. Because social constructs as facets of reality and objects of knowledge are not 'given' by nature, they must be constantly maintained and re-affirmed in order to persist. This process also introduces the possibility of change as well as persistence. Berger and Luckmann (1966) argue that when people interact, they do so with the understanding that their respective perceptions of reality are related, and as they act upon this understanding their common knowledge of reality becomes reinforced. Since this common-sense knowledge is negotiated by people, human typifications, significations and institutions come to be presented as part of an objective reality and taken for granted. Relevant for us is the involvement with this 'objective reality' by social scientists exploring the constructions of science and technology, in other words regarding them as mutable (see Latour and Woolgar, 1979; Bijkjer et al., 1987). Dake (1992) argues that the cultural contexts in which hazards are framed and debated and in which risk taking and risk perception occur is key, providing socially constructed myths about nature – systems of belief that are reshaped and internalized by persons, becoming part of their worldview and influencing their interpretation of natural phenomena. Risks then are issues that are defined, debated and acted upon by social members within a cultural context, the system characteristics of which must be explored to determine what a risk is. As Short (1984) notes these can help determine whether the hazard threatens the individual and/or the social fabric as well as the ideas of equity, fairness, transparency, legitimacy and trust. But as Lupton (1993) argues, risk discourse is often used to blame the victim, to displace the real reasons for ill health upon the individual and to express outrage at behaviour deemed socially unacceptable. 'Risk discourse' is not a neutral term, and culturally constructed judgments are used to evaluate risk situations and those within them. Risk constructions are therefore moral with, as we shall see, particular sciences trying to make them 'objective'. However, taken together with Sabatier's ideas about core beliefs, social constructions of risk may be very resistant to change even in the face of evidence that is compelling to other groups. This does not mean that we advocate for a retreat to relativism whereby any and all constructions of risk are legitimate but that only by understanding how risks are socially constructed can we ensure policy change is non-regressive, minimizing risks for those most in need of exposure reduction.

References

Adger, W.N. 2006. Vulnerability. *Global Environmental Change*, 16, 268–281.

Attorney General. 2002. *The Report of the Walkerton Inquiry*. Toronto: Queen's Printer.

Baxter, J., Eyles, J., & Elliott, S. 1999. 'Something happened': The relevance of the risk society for describing the siting process for a municipal landfill. *Geografiska Annaler: Series B, Human Geography*, 81(2), 91–109.

Beck, U. 1992. *Risk Society: Towards a New Modernity*. London: Sage.

Been, V. 1992. What's fairness got to do with it? Environmental justice and the siting of locally undesirable land uses. *Cornell Law Review*, 78, 1001.

Berger, P.L., & Luckmann, T. 1966. *The Social Construction of Reality: A Treatise in the Sociology of Knowledge*. Garden City: Anchor Books.

Biever, C. 2009. Katrina floods 'were government's fault'. *New Scientist*, 2707, 10.

Bijker, W., Hughes, T., & Pinch, T. 1987. *The Social Construction of Technological Systems: New Directions in the Sociology and History of Technology*. Cambridge: MIT Press.

Billings III, F.T. 2005. Cancer corridors and toxic terrors – is it safe to eat and drink? *Transactions of the American Clinical and Climatological Association*, 116, 115.

Bitektine, A. 2011. Toward a theory of social judgments of organizations: The case of legitimacy, reputation, and status. *Academy of Management Review*, 36(1), 151–179.

Blaikie, P., Cannon, T., Davis, I., & Wisner, B. 1994. *At Risk: Natural Hazards, People's Vulnerability, and Disasters*. London: Routledge.

Blodgett, A.D. 2006. An analysis of pollution and community advocacy in "Cancer Alley": Setting an example for the environmental justice movement in St. James Parish, Louisiana. *Local Environment*, 11(6), 647–661.

Brent, K. 2006. Gender, race, and perceived environmental risk: The "white male" effect in cancer alley, LA. *Risk*, 11(6), 453–478.

Bridges, J.W., & Bridges, O. 2004. Integrated risk assessment and endocrine disrupters. *Toxicology*, 205(1), 11–15.

Briggs, D. 2008. A framework for integrated environmental health impact assessment of systemic risks. *Environmental Health*, 7, 61.

Bullard, R. 1990. *Dumping in Dixie*. Boulder, CO: Westview Press.

CTV. 2011. *Ontario Pays 72m to Victims of Walkerton Tragedy*. Available at: http://www.ctvnews.ca/ontario-pays-72m-to-victims-of-walkerton-tragedy-1.646785#ixzz39X6ii7pV (Accessed 5 June 2015).

Cutter, S.L. 1996. Vulnerability to environmental hazards. *Progress in Human Geography*, 20, 529–539.

Dake, K. 1992. Myths of nature: Culture and the social construction of risk. *Journal of Social Issues*, 48(4), 21–37.

De Melo-Martín, I., & Meghani, Z. 2008. Beyond risk: A more realistic risk–benefit analysis of agricultural biotechnologies. *EMBO Reports*, 9(4), 302–306.

Douglas, M., & Wildavsky, A. 1982. *Risk and Culture: An Essay on the Selection of Technological and Environmental Dangers*. Berkeley: University of California Press.

Durmusoglu, E., Taspinar, F., & Karademir, A. 2010. Health risk assessment of BTEX emissions in the landfill environment. *Journal of Hazardous Materials*, 176(1), 870–877.

EPA. 2014. *NRC Risk Assessment Paradigm*. Available at: http://www2.epa.gov/fera/nrc-risk-assessment-paradigm (Accessed 5 June 2015).

EPA. 1995. *William K. Reilly Oral History Interview*. https://www.epa.gov/aboutepa/william-k-reilly-oral-history-interview (Accessed 5 June 2015).

Eyles, J., Taylor, S.M., Baxter, J., Sider, D., & Williams, D. 1993. The social construction of risk in a rural community: Responses of local residents to the 1990 Hagersville (Ontario) tire fire. *Risk Analysis*, 13(3), 281–290.

Furedi, F. 1997. *Culture of Fear: Risk-Taking and the Morality of Low Expectation*. London: Cassell.

Füssel, H.M. 2007. Vulnerability: A generally applicable conceptual framework for climate change research. *Global Environmental Change*, 17(2), 155–167.

Garg, A.X., Macnab, J., Clark, W., Ray, J.G., Marshall, J.K., Suri, R.S., . . . & Haynes, B. 2005. Long-term health sequelae following E. coli and campylobacter contamination of municipal water: Population sampling and assessing non-participation biases. *Canadian Journal of Public Health*, 96(2), 125–130.

Gartrell, C.D. 1987. Network approaches to social evaluation. *Annual Review of Sociology*, 13, 49–66.

German Advisory Council on Global Change. 2000. *New Strategies for Global Environmental Policy*. Berlin: WBGU.

Grothmann, T., & Patt, A. 2005. Adaptive capacity and human cognition: The process of individual adaptation to climate change. *Global Environmental Change*, 15, 199–213.

Hayes, M.V. 1992. On the epistemology of risk: Language, logic and social science. *Social Science & Medicine*, 35(4), 401–407.

Hirsch, R., Baxter, J., & Brown, C. 2010. The importance of skillful community leaders: Understanding municipal pesticide policy change in Calgary and Halifax. *Journal of Environmental Planning and Management*, 53(6), 743–757.

Jaynes, E. 2003. *Probability Theory: The Logic of Science*. Cambridge: Cambridge University Press.

Kasperson, J.X., & Kasperson, R.E. 2005. *The Social Contours of Risk*. London: Earthscan.

Kingdon, J. 1995. *Agendas, Alternatives, and Public Politics*. (2nd ed.) New York: Harper Collins.

Lacey, H. 2005. *Values in Science*. Lanham: Rowman & Littlefield.

Latour, B., & Woolgar, S. 1979. *Laboratory Life: The Construction of Scientific Facts*. Princeton, NJ: Princeton University Press.

Lupton, D. 1993. Risk as moral danger: The social and political functions of risk discourse in public health. *International Journal of Health Services*, 23(3), 425–435.

Lynch, K. 1984. *A Theory of Good City Form*. Cambridge, MA: MIT Press.

Marx, K. 1963. *The Machine in the Garden: Technology and the Pastoral Ideal in America*. Oxford: Oxford University Press.

Mindell, J., & Joffe, M. 2003. Health impact assessment in relation to other forms of impact assessment. *Journal of Public Health Medicine*, 25(2), 107–112.

Mohai, P., Lantz, P.M., Morenoff, J., House, J.S., & Mero, R.P. 2009. Racial and socioeconomic disparities in residential proximity to polluting industrial facilities: Evidence from the Americans' changing lives study. *American Journal of Public Health*, 99(Suppl 3), S649–S656.

Morello-Frosch, R., & Shenassa, E.D. 2006. The environmental "riskscape" and social inequality: Implications for explaining maternal and child health disparities. *Environmental Health Perspectives*, 114, 1150–1153.

Morello-Frosch, R., Zuk, M., Jerrett, M., Shamasunder, B., & Kyle, A.D. 2011. Understanding the cumulative impacts of inequalities in environmental health: Implications for policy. *Health Affairs*, 30(5), 879–887.

Morrow, B. 1999. Identifying and mapping community vulnerability. *Disasters*, 23(1), 1–18.

National Research Council. 1983. *Risk Assessment in the Federal Government: Managing the Process*. Washington, DC: National Academies Press.

National Research Council. 1996. *Understanding Risk: Informing Decisions in a Democratic Society*. Washington, DC: National Academies Press.

Nelkin, D. 1985. 'Introduction: Analyzing Risk.' in Nelkin, D. (ed.). *The Language of Risk*. Beverly Hills, CA: Sage.

New Orleans: Houses can be Rebuilt, but can Trust in Central Government? 2014. *The Guardian*. 27 January.

Oreskes, N., & Conway, E. 2010. *Merchants of Doubt*. New York: Bloomsbury Press.

Parr, J. 2005. Local water diversely known: Walkerton Ontario, 2000 and after. *Environment and Planning D: Society and Space*, 23(2), 251.

Perkel, C.N. 2002. *Well of Lies: The Walkerton Water Tragedy*. Toronto: McClelland & Stewart Ltd.

PHAC. 2006. *Integrated Risk Management*. Ottawa: PHAC.

Pidgeon, N., Simmons, P., & Henwood, K. 2006. 'Risk, Environment, and Technology.' in Taylor-Gooby, P. & Zinn, J. (eds.). *Risk in Social Science*. New York: Oxford University Press. pp 94–116.

Public Health Agency of Canada (PHAC). 2013. *Strategic Risk Communications Framework Within the Context of Health Canada and the PHAC's Integrated Risk Management*. Available at: http://www.phac-aspc.gc.ca/publicat/2007/risk-com/index-eng.php (Accessed 5 June 2015).

Ross, P.S., & Birnbaum, L.S. 2003. Integrated human and ecological risk assessment: A case study of persistent organic pollutants (POPs) in humans and wildlife. *Human and Ecological Risk Assessment*, 9(1), 303–324.

Royal Society. 1992. *Risk*. London: Royal Society.

Sabatier, P.A. 1988. An advocacy coalition framework of policy change and the role of policy-oriented learning therein. *Policy Sciences*, 21(2–3), 129–168.

Satterfield, T.A., Mertz, C.K., & Slovic, P. 2004. Discrimination, vulnerability, and justice in the face of risk. *Risk Analysis*, 24(1), 115–129.

Savadori, L., Savio, S., Nicotra, E., Rumiati, R., Finucane, M., & Slovic, P. 2004. Expert and public perception of risk from biotechnology. *Risk Analysis*, 24(5), 1289–1299.

Schlosberg, D. 2013. Theorising environmental justice: The expanding sphere of a discourse. *Environmental Politics*, 22(1), 37–55.

Short, J.F. 1984. The social fabric at risk: Toward the social transformation of risk analysis. *American Sociological Review*, 49(6), 711–725.

Tanner, C.M., Kamel, F., Ross, G., Hoppin, J.A., Goldman, S.M., Korell, M., . . . & Langston, J.W. 2011. Rotenone, paraquat, and Parkinson's disease. *Environmental Health Perspectives*, 119(6), 866–872.

Taubes, G., & Mann, C.C. 1995. Epidemiology faces its limits. *Science*, 269(5221), 164–169.

Taylor-Gooby, P., & Zinn, J. 2006. 'The Current Significance of Risk' in Taylor-Gooby, P. & Zinn, J. (eds.). *Risk in Social Science*. New York: Oxford University Press. pp 1–19.

Törnqvist, R., Jarsjö, J., & Karimov, B. 2011. Health risks from large-scale water pollution: Trends in Central Asia. *Environment International*, 37(2), 435–442.

Tsai, S.P., Cardarelli, K.M., Wendt, J.K., & Fraser, A.E. 2004. Mortality patterns among residents in Louisiana's industrial corridor, USA, 1970–99. *Occupational and Environmental Medicine*, 61(4), 295–304.

Turner II., B.L., Kasperson, R.E., Matson, P.A., McCarthy, J.J., Corell, R.W., Christensen, L., . . . & Schiller, A. 2003. A framework for vulnerability analysis in sustainability science. *Proceedings of the National Academy of Sciences*, 100, 8074–8079.

Ungar, S. 2001. Moral panic versus the risk society: The implications of the changing sites of social anxiety. *The British Journal of Sociology*, 52, 271–291.

Villani, A.C., Lemire, M., Thabane, M., Belisle, A., Geneau, G., Garg, A.X., . . . & Marshall, J.K. 2010. Genetic risk factors for post-infectious irritable bowel syndrome following a waterborne outbreak of gastroenteritis. *Gastroenterology*, 138(4), 1502–1513.

Vince, G. 2005. *Gauging the Health Crisis in Katrina's Wake*. Available at: https://www.newscientist.com/article/dn7959-gauging-the-health-crisis-in-katrinas-wake/ (Accessed 5 June 2015).

Weber, M., Driessen, P.P., Schueler, B.J., & Runhaar, H.A. 2013. Variation and stability in Dutch noise policy: An analysis of dominant advocacy coalitions. *Journal of Environmental Planning and Management*, 56(7), 953–981.

Whish-Wilson, P. 2002. The Aral Sea environmental health crisis. *Journal of Rural and Remote Environmental Health*, 1(2), 29–34.

Wisner, B., Blaikie, P., Cannon, T., & Davis, I. 2004. *At Risk: Natural Hazards, People's Vulnerability and Disasters*. New York: Routledge.

Xu, C. 2011. The fundamental institutions of China's reforms and development. *Journal of Economic Literature*, 49, 1076–1151.

3 How certain is the cost or benefit? Can it be made safe?

Introduction

In this chapter, we will examine how environmental health risk is considered and dealt with by experts. These experts are often quantitative scientists, trained in engineering, toxicology, epidemiology and economics. These activities are vital in a complex society in which there are considerable interactions between humans and physical and technical systems. As a result, codes of practice and health standards are created to protect workers and residents from too much risk. As we shall see, 'too much risk' is a fraught topic leading to considerable intellectual and political debate.

Risk and safety

Many concerns about the environment and its (usually adverse) effects on human health stem from the identification of hazardous events, activities and substances which threaten not only health but safety. A hazard can cause a threat to safety or have adverse consequences on someone in a workplace or environment. Many regulations and guidelines have been put in place to counteract these hazardous situations with some concerns that these may become overprotective or penalize a population because of specific dangers. Yet as Adams (2003) notes, nearly everyone has a propensity to take risks, although this varies among individuals and is influenced by the potential rewards of risk taking. Losses or costs are then a consequence of taking risks.

Much attention has been given to road and traffic safety such that most people are comfortable with a certain level of risk and aim to balance the rewards of risk taking against perceived hazards. So when seat belt use is mandatory and car design improves, the rewards of risk taking become more attractive and lead to a compensatory increase in risk taking (risk compensation), which may bring accident rates back to their original level (risk homoeostasis), or may produce a rearrangement of hazard with the new risk being transferred to others (risk displacement; Wilde, 1994; Adams, 1995).

The European Union has carried out much comparative work on road safety, with its data showing the relative hazards of different types of transport (see Figures 3.1, 3.2). Car occupants are 10 times more likely to die than those travelling in buses. But there is a good correspondence to traffic volumes and fatality rates. There is a geography to these risks to safety. In Britain, for example, two thirds of all road deaths occur on rural roads, especially when compared to the high-quality motorway network; single carriageways claim 80% of rural deaths and serious injuries, while 40% of rural car occupant casualties are in cars in collision with roadside objects, such as trees and poles. Cyclists of course have increased concerns over their safety. In Britain, they are eight times more likely to be injured than car users. Pedestrians fare more poorly with WHO (2013) reporting their being overrepresented in road traffic fatalities in Bangladesh, El Salvador, Ghana and the Republic of Korea, while they form a smaller proportion in the Netherlands, Thailand and the United States. In Britain, pedestrian deaths have declined 36% over a 10-year period (NAO, 2009).

Road calming measures (e.g., narrowing of streets), segregation of road users and pedestrians altering their travel behaviour are possible reasons for these declines. Cyclist deaths have, however, tended to increase. In the main, then, engineering solutions, abetted by regulatory change, have largely improved road safety.

Transport mode used by user	Fatalities per billion passenger kilometers
Airline passenger	0.101
Railway passenger	0.156
Bus/Coach occupant	0.433
Car occupant	4.450
Powered two-wheelers	52.593

Fatality risk ratios for transports	Airline passenger	Railway passenger	Bus/Coach occupant	Car occupant	Powered two-wheelers
Powered two-wheelers	520	337	121	12	1
Car occupant	44	28.5	10	1	
Bus/Coach occupant	4.3	2.8	1		
Railway passenger	1.5	1			
Airline passenger	1				

Figure 3.1 Fatality risk by transport mode

Source: Adapted from Eurostat (2012).

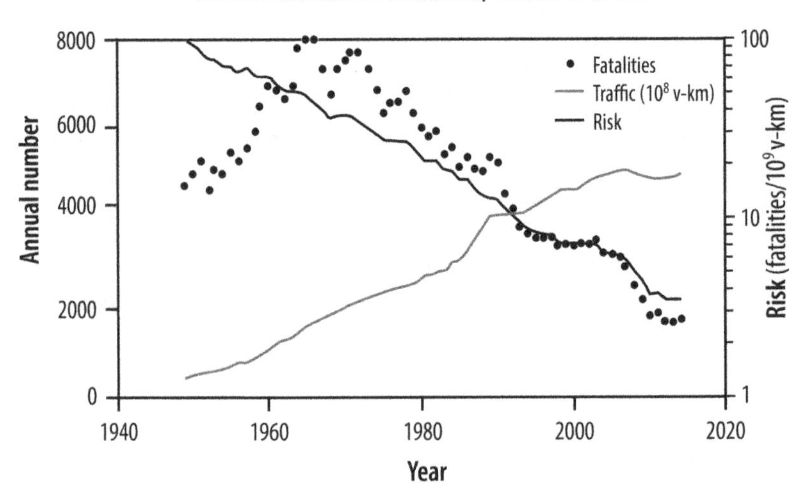

Figure 3.2 Road fatalities and car usage, Great Britain

Source: Heydecker (2015, personal communication). Courtesy of Dr. Benjamin Heydecker, Data from Department for Transport. 2015. Reported road casualties in Great Britain: annual report 2014. Presentation after Oppe, S. & Koornstra, M.J. 1990. 'A Mathematical Theory for Related Long Term Developments of Road Traffic and Safety'. in Koshi Masaki (ed.). *Transportation and Traffic Theory: Proceedings of the Eleventh International Symposium on Transportation and Traffic Theory*, July 18–20, 1990, in Yokohama, Japan. New York: Elsevier, pp 113–132.

Safety, acceptable risk and strength of associations

In many ways, epidemiology and toxicology have worked similarly using expert systems to enhance safety and reduce hazard. For example, a key contribution of toxicology for understanding chemical safety is the relationship between dose and its effects on the exposed organism. The chief criterion regarding the toxicity of a chemical is the dose, or the amount of exposure to the substance. Epidemiology is the science that studies the patterns, causes and effects of health and disease conditions in defined populations (see Chapter 2). It is the cornerstone of public health. Both apply scientific procedures to determine what is a safe dose or acceptable risk and how a might substance affect an organism, individual or population.

Determining an acceptable risk is problematic. Hunter and Fewtrell (2001) outline the possible definitions. A risk is acceptable when it falls below an informed defined probability or some level that is already tolerated or below an arbitrary defined attributable fraction of total disease burden in the community. As Chapter 2 noted, these definitions are culturally determined and accepted. An economic component can be added – the cost of reducing the risk would exceed the costs saved, the risk would exceed the costs saved when the 'costs of suffering' are also factored in or the opportunity costs are too high – that is, resources would be better spent on other, more pressing public health problems. It can also be political – public health professionals say it is acceptable or the general public say it is acceptable (or, more likely, do not say it is not) or politicians say it is acceptable. In this chapter we deal with the scientific and economic arguments around safety, cost and benefit. The other approaches are examined in later chapters.

It is often quite difficult to calculate threshold and sub-threshold doses for substances. A whole series of safety factors (uncertainty ones) is used, relying on animal studies, modifications for such factors as inter-human variation in susceptibility, limited data and calculations of no or lowest observed adverse effect (NOEL and LOEL; see Chapter 1). Thus professional assessments are used to make conservative estimates of safe dose. For most of these uncertainties, a 10-fold increase, or safety factor, is included so if the available evidence points to a 1 in 100,000 risk of the threshold dose being hazardous, 1 in 1 million is applied for a safe dose (see Dourson et al., 1996).

Despite these scientific responses to safety concerns about toxic substances, much debate still occurs about what happens at low doses and whether the dose-response gradient always takes the same, albeit conservative, form (see Figure 3.3 for a schematic of various methods). Even if there are few problems, correct assessment of a toxin causing health effects can take a long time. Methylmercury provides a telling case, with more than 120 years required from the first fatality to international action (see Grandjean et al., 2010).

Similar scientific uncertainties were found in the development of occupational health legislation. Early studies of the variations in mortality between towns and counties show that cancer deaths are affected by the presence of etiological factors in the environment. Gardner et al. (1982) found that high mortality ratios for pleural mesothelioma were particularly high in dockyards and ports which had at the time high uses of asbestos, particularly crocidelite, used in shipbuilding and

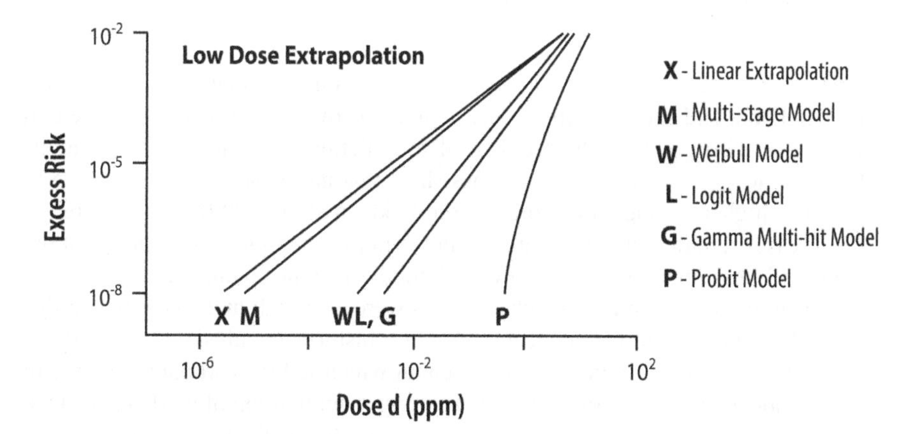

Figure 3.3 Schematic of various low-dose extrapolation methods

Source: WHO (2000), with kind permission of the World Health Organization.

Table 3.1 Important early warning about and recognition of methylmercury (MeHg) toxicity

Year(s)	*Event*
1865	First published record of fatal occupational MeHg poisoning
1887	First experimental studies on MeHg toxicity
1930	Report on organic mercury poisoning in acetaldehyde production workers
1940–1954	Poisoning cases in workers at MeHg fungicide production plants
1952	First report on development of MeHg neurotoxicity in two infants
1956	Discovery of a seafood-related disease of unknown origin in Minamata, Japan
1959	Studies on MeHg toxicity in cats suppressed by the polluting company
1967	Demonstration of mercury methylation in sediments
1968	Official acknowledgement of MeHg as cause of Minamata disease
1955–1972	Occurrence of poisoning epidemics from the use of MeHg-treated seed grain for cooking, decline in exposed wildlife populations
1973	Report on dose-response relationship in adults from Iraqi data
1986	First epidemiology report on adverse effects in children related to maternal fish intake in New Zealand
1997	Confirmation from prospective study in the Faroe Islands on adverse effects in children from MeHg in maternal seafood intake during pregnancy
1998	White House workshop of 30 scientists identifies uncertainties in evidence
2000	NRC supports exposure limit of 0.1ug/kg per day
2003	Updated JECFA exposure limit of 1.6ug/kg per week
2004	European Union expert committee recommends that exposures be minimized
2005	European Union decides on a ban on mercury exports
2009	International agreement on controlling mercury pollution

Source: Adapted from Grandjean et al. (2010).

repair. High rates of nose and sinus cancers were found in men working in furniture and boot and shoe industries. Nasal cancer is rare, so estimates are based on few cases. This points to the problem of formulating guidelines and regulations based on few cases. But recent reviews show that nasal cancer incidence is 5 to 10 times higher among shoe workers (see Pukkala et al., 2009). While there are fewer cases reported, this may suggest not better occupational conditions but the export of these manufactures to low- and middle-income countries.

In Gardner et al.'s study, bladder cancers were also high in areas in which dye and rubber industries were concentrated. In a historical analysis, Morris (1975) points to an epidemiological detective chain which led to workplace reform. In 1895 bladder cancer was first described among a small group of workers engaged in the manufacture of magenta from aniline, which was then incorrectly seen as the most likely agent. But political events changed the work picture as during the First World War, German supplies of dyestuffs ceased and their manufacture in Britain increased. Yet at this time the health problem could not be connected to the environmental risk, as the population-at-risk was small and classified in the occupational mortality statistics with other groups. Classifications might not always be able to assist in identifying associations and in recognizing a risk. Eventually careful study revealed the risk of ranging between 1 in 5 to 10 and for one small group, distilling β-naphthylamine one in one. In 1948 this product and x-naphthylamine and benzidine were found to be the etiological agents. In 1952, its manufacture was halted and screening introduced. Yet cases of this cancer are still found among rubber workers.

In these historical examples, we notice how rare diseases and classifications – the latter being a product of expert systems – do not always decrease uncertainty in assessing risk. Environmental epidemiology and health risk and impact assessment have long grappled with uncertainty in data and their relationships, particularly in analysis, less so in problem formulation and risk communication (see Briggs et al., 2009). Yet in trying to determine associations between environment and health, uncertainty is largely seen in scientific terms with respect to information uncertainty, limits of available analytic tools, complexity and indeterminacy of environmental and human systems and the role of value judgments in science itself (see Lemons et al., 1997). So for many, more and more detailed science will largely provide answers to the costs and benefits of an activity or event. But as Jamieson (1996) notes, seeing uncertainty merely as an objective quantity the value of which can be reduced is only part of the story. Uncertainty may be a disguise for fallibility – the association is wrong – which is itself compounded by the complexity of environment–health relations whereby more powerful analytic tools may improve measurement but not reduce uncertainty. Furthermore, all knowledge claims, including those of science, use categories or classifications to provide clarity for understanding. But categories or classes depend on the evidence available, the length of time data has been collected and so on. We will tackle these issues in later sections.

The ubiquity of uncertainty has meant that attempts have been made to specify the minimal conditions necessary to associate an exposure with an outcome. Bradford Hill (1965) established the most widely used scheme identifying strength (see Chapter 1). However, in some cases, the mere presence of the factor can trigger the effect. In other cases, an inverse proportion is observed: greater exposure leads to lower incidence (these are concerns of the dose-response gradient); plausibility: a plausible, biological mechanism between cause and effect is helpful but limited by current knowledge); coherence: between epidemiological and laboratory findings increases the likelihood of an effect (but application of safety factors may disguise this relationship); and finally experiment and analogy (the search for associations in similar situations). Grant (2009) used ecological data to demonstrate an association between solar UVB and vitamin D and reduced cancer risk for certain types, excluding skin cancer, satisfying these criteria.

Howick et al. (2009) update Hill's postulates by suggesting that associations should be derived from direct evidence from studies (randomized or non-randomized) that a probabilistic association between intervention or exposure and outcome is causal and not spurious, mechanistic evidence for the alleged causal process that connects the intervention and the outcome and parallel evidence that supports the causal hypothesis suggested in a study, with related studies that have similar results. Such an approach leads to a careful examination of all potential evidence of an association, including case reports, anecdotes and countervailing findings. For example, non-monotonic responses must be considered (see Figure 3.4). These are relevant in the investigation of the effects of endocrine disruptors, making a low observed adverse effect hard to calculate. A U-shaped dose response has been found for bisphenol A with low and high doses resulting in an oestrogen response (see Welshons et al., 2006). Endocrine disruption is a concern as chemicals can interfere with the endocrine (or hormone systems) in mammals. These disruptions can cause cancerous tumours, birth defects and other developmental disorders. Any system in the body controlled by hormones can be derailed by these disruptors. Specifically, endocrine disruptors may be associated with the development of learning disabilities, severe attention deficit disorder, cognitive and brain development problems, deformations of the body (including limbs), breast cancer, prostate cancer, thyroid and other cancers and sexual development problems such as feminizing of males or masculinizing effects on females. Many substances that potentially have these effects, such as polychlorinated biphenyls (PCBs), dichlorodiphenyltrichloroethane (DDT) and persistent organic pollutants (POPs) have been banned and body burden has declined. Yet safety remains compromised.

Such chemicals as atrazine, widely used on corn crops in the United States and a pervasive drinking water contaminant linked to breast tumours, delayed puberty and prostate inflammation in animals and phthalates found in some soft toys, flooring, medical equipment, cosmetics and air fresheners and implicated in the rise of birth defects of the male reproductive system and possibly the reproductive system of infants are still widely used. Their ubiquity in many products with

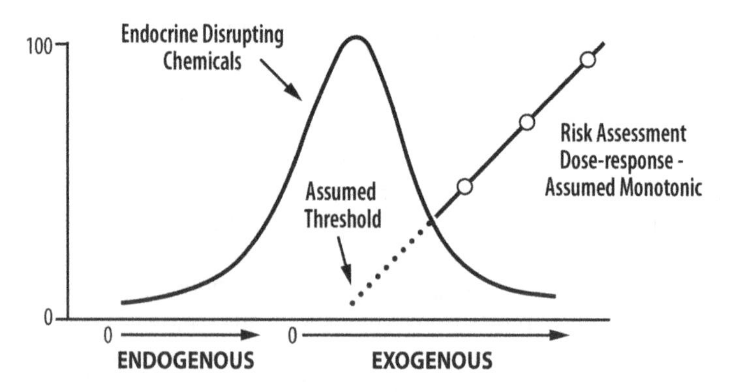

Figure 3.4 Dose-response curve types

Source: NIEHS (2012).

industrial support and lifestyle implications means that concentration levels have been reduced. Despite the threat of endocrine disruptors to human development, the scientific argument for a ban has to be certain. There needs to be very strong direct evidence of an exposure–outcome link and given all the factors that can lead to, say, a birth defect or reduced sperm count, such a link is extremely difficult to confirm.

Scientific uncertainty: False positives and false negatives

So with chemical exposures, scientific uncertainty often remains, making policy action problematic. As Wildavsky (1997, p. 36) says, science in the world of political action rather than just knowledge production enters "the strange world of weak causes and infinitesimal effects". With the power of citizen groups, substances, especially those with only a specialized constituency, often get banned, while those that affect a large portion of everyday life do not. For example, saccharin was investigated for human health impacts as past warnings were ignored (see Priebe and Kauffman, 1980). But scientific studies and industrial and political activity have shown that it is relatively 'safe' for human consumption, and it has subsequently been removed from banned substance lists (see California Office of Environmental Health Hazard Assessment, 2001). Different approaches were applied to dioxin, which became known as the most dangerous chemical. As is usual, health effect depends on dose. Yet trace amounts in road dust in Times Beach, Missouri, led to the town being abandoned and residents relocated. The cleanup of the town and other sites with dioxin cost $110 million. At Times Beach, the likely effects of minute doses were lost in an avalanche of hype and misreporting. No attention was paid to the psychological stress of residents (see Wildavsky, 1997). The area has been turned into a state park considered safe for human activity, although there remains public doubt of low test results ("Thirty Years After Dioxin Disaster," 2012). Dioxin, a by-product of industrial processes, remains a concern, although retrofitting incinerators and other technical solutions have reduced its production. Its potential broad health impacts (cancer, neurodevelopmental and endocrine disorders) have led to the introduction of a safety factor of 10, reducing tolerable daily intake from 10 pg/kg of body weight to 1 to 4 pg/kg (see van Leeuwen et al., 2000).

Science in politics intensifies the impacts of uncertainty. This may be compounded by the ways in which science operates to produce fallible results. False positives – a type I error leading to the conclusion that a supposed effect or relationship exists when in fact it does not – point to an association being true in a scientific sense when it is in fact false. Many people will recognize false positive/negative from pregnancy testing. In environmental health, the purpose of much statistical testing is to try and avoid these errors although underpowered or small samples may increase the likelihood of their appearance.

False positives can result in unnecessary public concern and even regulatory action (see Hansen et al., 2007). For example, the swine flu of 1976 led to a mass inoculation program which cost more than US$100 million and led to several deaths associated with the vaccine (see Hansen and Tickner, 2011). And while as Blair et al. (2009) note policy must be formulated on imperfect evidence, which may include false positives, there remains a concern about associations between environmental exposures and health impacts at the local level. Even with respect to the more closed environment of the workplace, Swaen et al. (2001) identified 150 true positive and 75 false positive studies utilizing scientific assessments about types of research design and kinds of evidence used. Similar concerns can be found in studies of residential environments where sample size (now too large) and exposure zone (potential redefinition) might result in false positive association between density and proximity of natural gas wells within a 10-mile radius of maternal residence and prevalence of congenital heart defects and possibly neural tube defects (see McKenzie et al., 2014). But correction through multiple analyses to reduce the likelihood of false positives can lead to all associations disappearing, as found in a recent study of asthma genes and industries in rural areas (see Morin et al., 2012).

False negatives are scientific claims that an effect which is really present is not there. So if a false positive gives a false alarm, a false negative gives no alarm at all. False negatives have been found in all screening programs, irrespective of the quality of the service. They can delay the detection of breast, cervical and prostate cancer with potential psychological and legal implications (see Petticrew et al., 2000). Some studies have found that screening for such microorganisms as Giardia misses their presence (see Harwood et al., 2005). Brown (1992, p. 274) suggests that a key value judgment of the science of environment and health is that "epidemiologists prefer false negatives to false positives – that is, they would prefer to claim falsely that an association between variables does not exist when it does than to claim an association when there is none." But direct evidence even if experiential or anecdotal can be used to question false negatives. As Paigan notes about the toxic contamination at Love Canal,

> I needed a 95 percent certainty before I was convinced of a result. But seeing this rigorously applied in a situation where the consequences of an error meant that pregnancies were resulting in miscarriages, stillbirths, and children with medical problems, I realized I was making a value judgment.
>
> (quoted in Brown, 1992, p. 274)

Scientific uncertainty is not reduced by such judgments or claims. Aronson (1984) identified two scientific knowledge claims. Cognitive claims aim to convert experimental observations, hypotheses and theories into publicly accredited factual knowledge – a public view of science as a truth maker. Interpretive claims establish the broader implications of the research findings for a non-specialist audience. Interpretive claims implicitly ask the audiences to certify the usefulness of the research which is based on the cognitive claims. For example a . . . cognitive claim . . . while an interpretive claim by a specific scientist – whether or not they conducted the original research – might be . . . One type of interpretive claim is technical in which researchers act as scientific advisers to industry and government. This often involves the evaluation of risks posed by controversial technologies or events. Second, cultural interpretive claims attempt to develop ideological support both for expenditures on scientific research and for the autonomy of science by public speeches, op-eds and so forth. Finally, social problem interpretive claims assert the existence of a social problem that a particular scientific specialty is uniquely equipped to solve. Interpretive claims get the public and funders on side for scientific practice, this often requiring definitive statements whatever the issues of small dose, small sample size and multi-causation of an adverse health outcome. Is the scientist a modern equivalent of a priest? Is s/he infallible? As with religious and other leaders and the ways of understanding they represent have been critically analyzed through the ages, various publics have become sceptical and learned ways to legitimately question scientists and science.

Scientific fallibility: What models are employed?

> In order to observe anything, in order to 'collect data', one must have some notion – no matter how primitive and preliminary – of the particular experiences one intends to relate to one another. It is, obviously, these experiences that one will be looking for. In order to find them, one necessarily assimilates and disregards all sorts of differences in individual observations. The longer this goes on successfully and the more often the model one has constructed proves useful, the stronger becomes the belief that one has discovered a real connection, if not a Law of Nature. And once that belief has been established, there is a powerful resistance against any suggestion of change.
>
> (von Glasersfeld, 1987, p. 9)

Such models create a conservatism in science. The criteria established by Hill and others support this so scientific proof is not lightly claimed. This may however create public and legal difficulties. For example, in Woburn, Mass., a childhood leukemia cluster seemed to appear in the late 1960s which was associated in residents' minds with chemical pollution of local wells with trichloroethylene, an industrial solvent used by two companies in the area. Establishing these connections required careful attention of likely associations with other causes of leukemia, identification of carcinogenic pathways, family behaviour and migration and the history of industrial development in the area. These requirements led CDC to conclude the evidence available meant no firm conclusion that the companies and their pollution were the culprits beyond scientific doubt was possible, and the Mass. Public Health Department concluded that leukemia rates had begun to rise before the problematic wells were constructed. It took citizen science, legal involvement and academic research to bring the issue to the courts, which rely more on probable cause than on scientific tenets. Some company culpability was found, but out-of-court settlement meant no legal wrong-doing was formally identified. Later scientific work on Woburn did find a strong association between prenatal exposure and leukemia risk but no link between childhood exposure and risk (see Costas et al., 2002), findings not consistent with other studies. Burden of proof remains problematic whereby nearby residents are meant to prove harm instead of polluters demonstrating safety (see Fuchs, 1996). That is, with the same scientific data we would come to very different political/legal conclusions if uncertainty favoured eliminating potentially harmful chemicals.

Cancer clusters remain a significant scientific and public concern. CDC (2013) defines a cancer cluster as a greater-than-expected number of cancer cases that occurs within a group of people in a geographic area over a defined period of time: the number of observed cases must be greater than one typically would observe in a similar setting (e.g., in a cohort of a similar population size and within demographic characteristics) and depends on a comparison with the incidence of cancer cases seen normally in the population at issue or in a similar community; the cancer cases must be all of the same type, unless an exposure is linked to more than one type; the population in which the cancer cases are occurring must be defined by its demographic factors (e.g., race/ethnicity, age, and sex), within a carefully designated geographic area and within a time period which affects both the total cases observed and the calculation of the expected incidence of cancer in the population. Discovering scientific associations is problematic. Goodman et al. (2012) analyzed 567 cancer cluster investigations in the U.S. since 1990. Only one was identified as having a strong scientific linkage between pleural cancer and shipyard employment in South Carolina. Residential studies have all proved to be inconclusive. There have therefore been attempts to improve the consistency and accuracy of scientific measurement. In the short term though, potential clusters raise serious questions about whether a more cautionary approach to the use of substances and technologies should be the general political/value principle that guides decision making in such cases. From the point of view of science though, improving accuracy is proving difficult; thus it may be problematic for dealing with such environmental and health problems in the short term.

Improving measurement of exposure

Exposure assessment measures the magnitude, frequency and duration of an exposure to an agent, with other characteristics of the population exposed. Ideally, it describes the sources, pathways and routes and the uncertainties in the assessment. Inhalation, ingestion and dermal absorption are routes of exposure. For example, there is concern about ensuring full likely exposure to trihalomethanes (THMs), a by-product of chlorinated water disinfection interacting with organic matter. THMs have been associated with colon and bladder cancer risk. Nazir and Khan (2006) found that inhalation during showering significantly increases risk.

More detailed science about the impact of exposures has come through the development of biomarkers – things we can measure in the human body to indicate exposure to substances (e.g., metabolites in urine).

> The use of biomarkers in basic and clinical research as well as in clinical practice has become so commonplace that their presence as primary endpoints in

> clinical trials is now accepted almost without question . . . In many cases, however, the "validity" of biomarkers is assumed where, in fact, it should continue to be evaluated and reevaluated.
>
> (see Strimbu and Tavel, 2010)

WHO (1993) defines a biomarker as "almost any measurement reflecting an interaction between a biological system and a potential hazard, which may be chemical, physical, or biological. The measured response may be functional and physiological, biochemical at the cellular level, or a molecular interaction". There remain doubts about the accuracy and validity of particular biomarkers. Their relevance must always be viewed as transitional, and ability to measure is not necessarily demonstration of an association. Of particular relevance are biomarkers of exposure which try and establish a potential toxic dose internally rather than through proxy external indicators in the environment or through questionnaire. Thus it measures a "biologically effective dose" indicating "the amount of toxin or chemical measured in the target organ or its surrogate" (Mayeux, 2004, p. 184). This can be measured through a number of body fluids. Some chemicals such as halogenated hydrocarbons are stored in adipose tissue, but others, such as organophosphate pesticides, are better measured in blood or urine.

The story of lead: Improving measurement, reducing uncertainty and creating health benefits

Lead exposure is an excellent example. A history of lead exposure can be strengthened by measurement of lead in the environment, but the best indication of the dose of exposure may be determined in blood and tissues (hair, nails, teeth; Figure 3.5). Sakai (2000) has noted the importance of measuring biomarkers of exposure, effect and susceptibility for lead to understand the impact of tissue damage. Exposure to high levels of lead has been known for decades. Contention developed over chronic low levels. Needleman and Gatsonis (1990) note that disagreement resulted from (1) selecting adequate markers of exposure or internal dose, measuring outcome with instruments of adequate sensitivity, (2) identifying, measuring, and controlling for factors that might confound the lead effect and (3) recruiting and testing a sample large enough to provide adequate statistical power to detect a small effect and designing a study that avoids biases in sample selection. Lead is a stable metal and was used in building construction, lead-acid batteries, bullets and shot, weights, as part of solders, pewters and fusible alloys and as a radiation shield. Despite being considered hazardous, it continued to be widely used in the 20th century in paint and as an additive to gasoline. Banning its use in paint began in the early 1900s but was not instituted in the U.S. till the 1970s. The research on the impact of low-level lead exposure was challenged by industry as it challenged their view that average lead levels were safe (see Figure 3.6). Court proceedings and legal challenges were instituted until simple associative analysis between blood levels and previous gasoline use were shown.

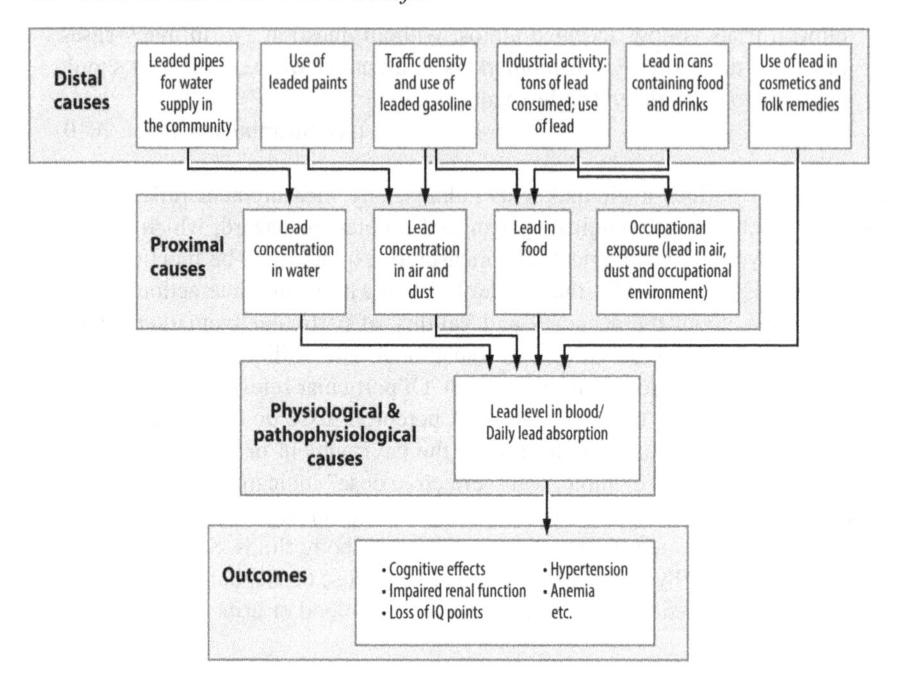

Figure 3.5 Framework of causes and outcomes for lead exposure

Source: WHO (2000), with kind permission of the World Health Organization.

Lead removal from gasoline required technical changes in car manufacture and contentious debate over whether this could be done and at what cost. But Japan banned lead from gasoline in 1986, Canada in 1990 and the U.S. in 1996. The European Union had debate on levels, but countries reduced and/or banned lead from the mid-1990s (see World Bank, 1998). Slowly middle-income countries (MICs) and low-income countries (LICs) have adopted similar regulations with only a handful of countries still using leaded gasoline. Residual lead continues to remain a problem. Jacobs et al. (2002) point to the continuing lead exposure in poor neighbourhoods with old lead-based paint and lead exposure in soil. These conditions affect children and their cognitive functions in particular. In recent analyses of lead concentrations in household dust, associations were found with income, race/ethnicity, floor surface or condition and year of construction. Furthermore, before its removal from gasoline, exposure disparities would likely have been linked to roadway proximity, but present exposures are more strongly associated with housing age and quality (see Adamkiewicz et al., 2011).

Costing environmental hazards to health

So far we have examined the costs of environmental hazards in terms of the risks they pose and the costs of managing the negative health outcomes within the

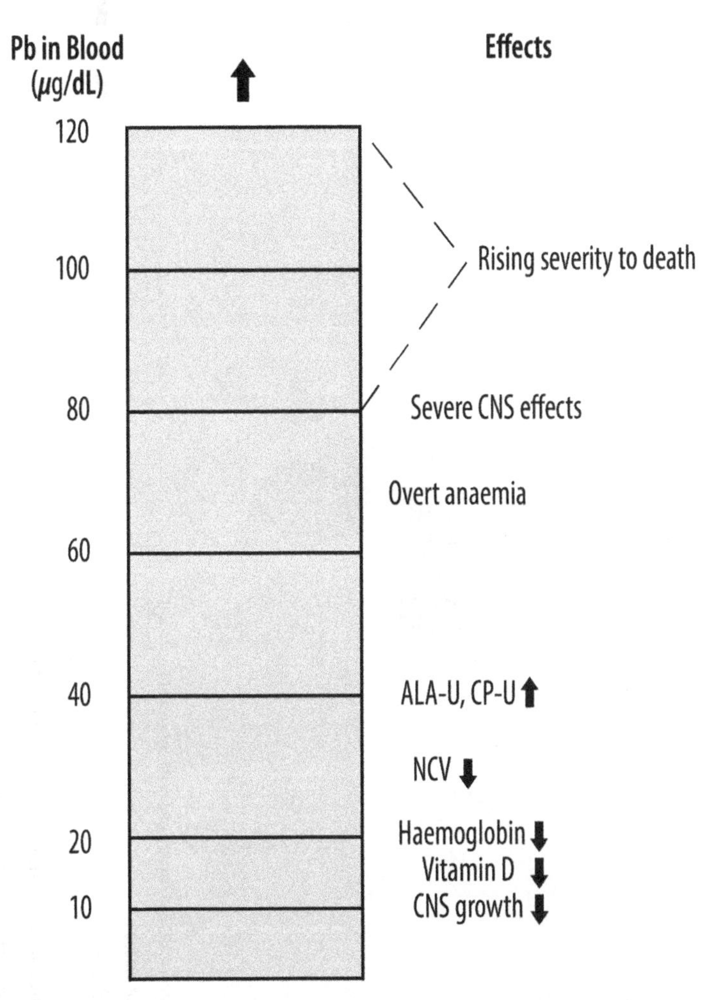

Figure 3.6 Lowest observable adverse effects levels of lead exposure

Source: Tong et al. (2000), with kind permission of the World Health Organization.

current health system – this is called the burden of disease. This can be carried out in terms of adverse outcome such as maternal mortality, which has strong environmental and access-to-care components (see Figure 3.7).

It can also be carried out by trying to calculate the burden of disease determined by specific hazards. There has been much interest on the contribution of different

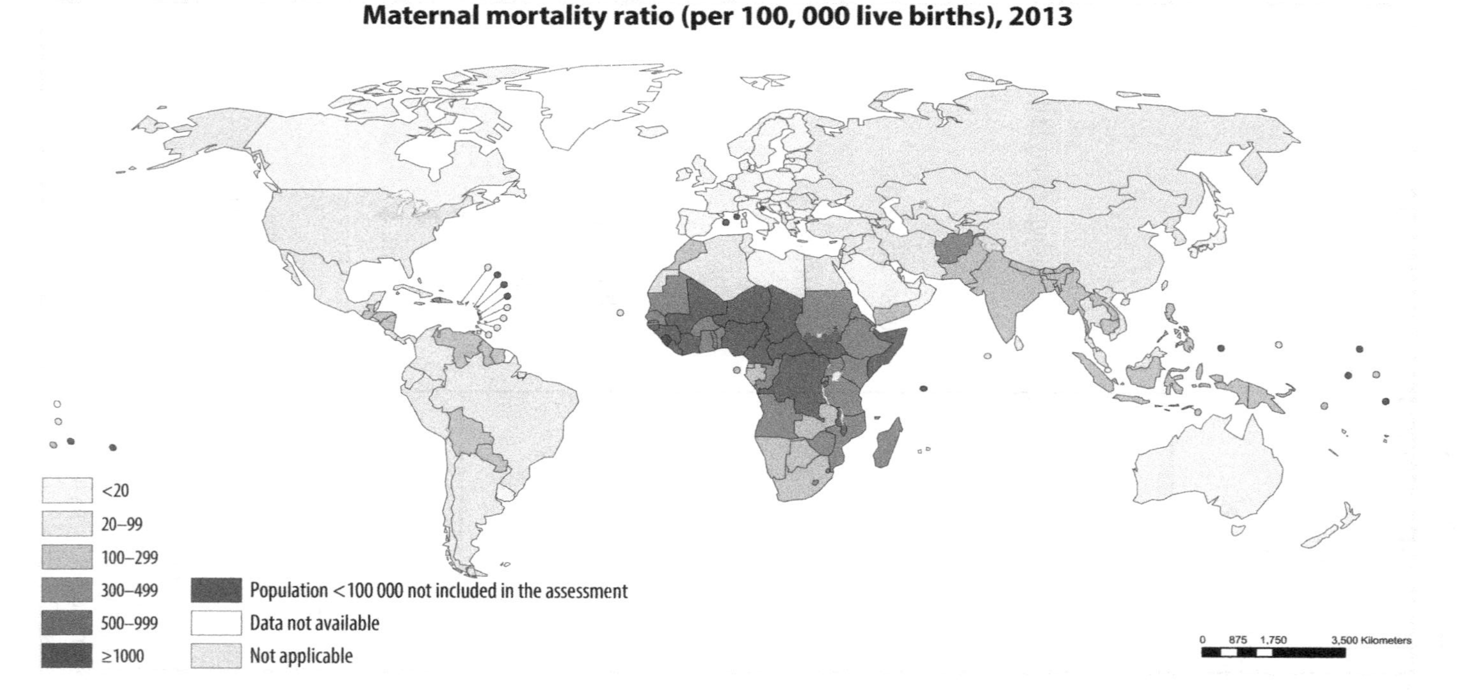

Figure 3.7 Countries by category according to maternal mortality ratio (MMR, death per 100,000 live births)

Source: WHO (2014), with kind permission from the World Health Organization.

pollutants to disease burden. As the European example shows, particulate matter is an overwhelming contributor. Hänninen et al. conclude (see Figure 3.8.),

> EBD [environmental burden of disease] was estimated for nine environmental risk factors (benzene, dioxins, formaldehyde, SHS [second hand smoke], lead, traffic noise, PM2.5, ozone, and radon) in six countries. The highest overall public health impact was estimated for ambient fine particles (PM2.5; annually 4,500–10,000 nondiscounted DALYs (Disability-Adjusted Life Years)/million in the six participating countries) followed by second hand smoke (SHS) (600–1,200), traffic noise (400–1,500), and radon (450–1,100). Medium impacts were estimated for lead, dioxins, and ozone. Lowest impacts were estimated for benzene and formaldehyde. The relative ranking of the risk factors was relatively robust under the uncertainties examined.
>
> (2014, p. 445)

To identify further the costs of environmental hazards on health, disability-adjusted life year (DALY), a measure of overall disease burden, expressed as the number of years lost due to ill-health, disability or early death can be used. DALYS can be weighted for age and can take into account living with disabilities and premature mortality. DALY summarizes burden of disease among populations combining mortality and morbidity measures, thus assessing non-fatal outcomes as part of the burden. It needs a standard means of weighing different types and severities of disability (see Thacker et al., 2006). WHO (2014) has calculated DALY by region and country. The 10 countries with the highest age-standardized DALY rates are in Sub-Saharan Africa. Sierra Leone bears the highest burden of disease, despite having experienced a decline in age-standardized DALY rate from 157,900 to 117,700 per 100,000 population between 2000 and 2012.

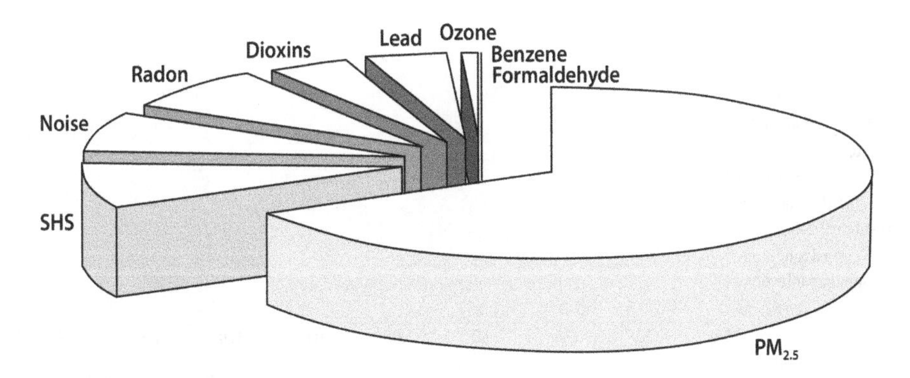

Figure 3.8 Relative contributions of risk factors to the estimated burden of disease attributed to them, averaged over six countries

Source: European Environmental Agency (2011), Opasnet.

In the United States, for example, heart disease accounts for the largest fraction of lost DALY for both men and women. HIV/AIDS, alcohol abuse and depression are also important causes of lost DALY, with depression being the second leading cause for women (see Thacker et al., 2006). Different diseases result in an altered picture for low-income countries, such as Laos (see Figure 3.9).

Globally, there is a different picture:

> In 1990, 47% of DALYs worldwide were from communicable, maternal, neonatal, and nutritional disorders, 43% from non-communicable diseases, and 10% from injuries. By 2010, this had shifted to 35%, 54%, and 11%, respectively. Ischaemic heart disease was the leading cause of DALYs worldwide in 2010 (up from fourth rank in 1990, increasing by 29%), followed by lower respiratory infections (top rank in 1990; 44% decline in DALYs), stroke (fifth in 1990; 19% increase), diarrhoeal diseases (second in 1990; 51% decrease), and HIV/AIDS (33rd in 1990; 351% increase). Major depressive disorder increased from 15th to 11th rank (37% increase) and road injury from 12th to 10th rank (34% increase). Substantial heterogeneity exists in rankings of leading causes of disease burden among regions.
>
> (see Murray et al., 2012, p. 2198)

Attempts have also been made to take in the financial or economic cost of environmental hazards on health, a difficult task because of the wide range of factors that lead to or can change cost, for example increases in treatment prevalence, rising costs of cases, greater access to services and so on. All health hazards lead of course to medical costs and individual and societal ones. OECD (2014) uses the value of a statistical life, determined by how much an individual is willing to pay to reduce death from a particular risk to estimate the costs of air pollution from road transport. These approaches privilege mortality costs over morbidity ones and attempt to take into account resource or direct costs (medical, care),

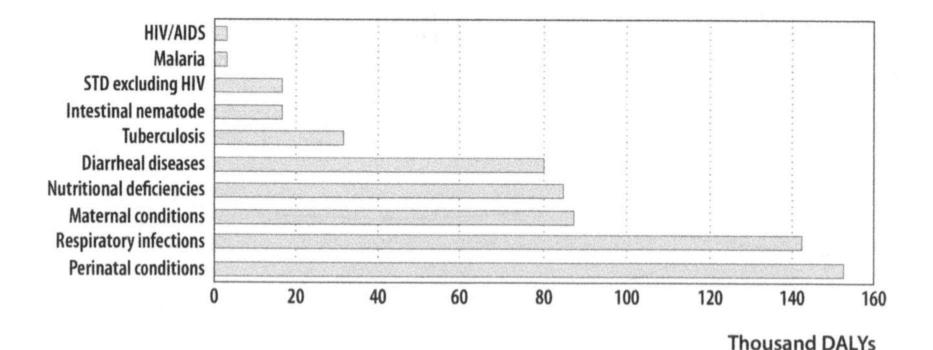

Figure 3.9 DALYs by disease type in Laos

Source: Longfield et al. 2013 licensed under CC BY 4.0 from BMC Public Health.

opportunity or indirect ones (loss of productivity, leisure time and disutility ones (pain, discomfort). It is a difficult analytic process, but OECD (2014) concludes that the health cost of air pollution in all OECD countries was around US$1.7 trillion in 2010 (about the size of the Canadian economy), with more than 50% arising from road transport. About two thirds of OECD countries show a decline in premature mortality from air pollution and therefore in relative costs, but some, including the U.S., Canada, Japan, Turkey and South Korea, show increased costs between 2005 and 2010, as do the industrializing powers of China and India.

There are local examples of the economic costs of environmental hazards with respect to health. For Ontario, the Ontario Medical Association (OMA) estimated that there were 5,800 premature deaths due to air pollution in 2005. With an estimated 16,000 hospital admissions and 60,000 emergency room visits, the OMA calculated the costs as being $374 million in lost productivity and work time, $507 million in direct health care costs, $537 million in pain and suffering due to non-fatal illness and $6.4 billion in social welfare loss due to premature death from air pollution. Yet recent studies in Hamilton, Ontario, show that with a changing industrial structure and improved technical applications to manufacturing plants, there has been a reduction in the costs of air pollution to health and the economy in general between 2003 and 2012 (see SENES, 2012). In other parts of the world, Wang and Mullahy (2006) note that low wage levels and the costs of reducing air pollution make clear air a 'luxury'. In Sweden, willingness to pay for improved air quality increased with income, wealth and education; it was larger for men, members of environmental organizations, people living in big cities (which are on average more polluted) and people who own their house or apartment. It was lower for retired people (see Carlsson and Johansson-Stenman, 2000). Yet while these investigations of the economic cost of health-affecting hazards are interesting and focus debate, they tend to be contentious because of methodological issues about sample size, omitted variables, question formulation, the dimensions of value in a statistical life and setting hedonistic prices (i.e., the value of life is imputed from expressed preferences that do not specifically ask the value of a life) in willingness-to-pay studies. As we have noted before, contentious policy demands more detailed scientific evidence to improve policy formulation. Yet that scientific evidence itself can be limited in numerous ways. This may well lead to scientific findings themselves becoming contentious and challenged politically and legally.

There is also a public argument that certain activities and technologies produce many positive benefits as well as these externalities. For example, the use of pesticides may reduce populations of disease-carrying mosquitos or increase the quantity of a crop produced and a farmer's income but create costs for others if pesticide residue seeps into ground- or drinking-water sources. It is necessary but often very difficult to establish all the interactions the may lead to all individual and social costs and benefits as well as the derivation of positive externalities in which individual gain comes at a societal cost – a tragedy-of-the-commons situation. Most of these methods are, in fact, types of cost-benefit analysis in which precise detail is placed on the value of a life or the role of the individual as a consumer rather than as a citizen. Willingness to pay is a way of trying to price that which is not costable, a statistical life is not the value of a life but the estimated cost of reducing certain risks, discounting the future removes later concerns by imposing a calculated investment income and citizens are diminished to their economic choices, which will vary according to their resources. The logic of cost-benefit was expressed in a 1991 World Bank memo –

> the measurements of the costs of health impairing pollution depend on the foregone earnings from increased morbidity and mortality. From this point of view a given amount of health impairing pollution should be done in the country with the lowest cost, which will be the country with the lowest wages. I think the economic logic behind dumping a load of toxic waste in the lowest wage country is impeccable and we should face up to that.
>
> (Summers, quoted by Ackerman and Heinzerling, 2002, p. 1574)

Such comments – soon disowned – become contentious and raise public concerns about international risk management and environmental justice.

References

Ackerman, F., & Heinzerling, L. 2002. Pricing the priceless: Cost-benefit analysis of environmental protection. *University of Pennsylvania Law Review*, 150(5), 1553–1584.

Adamkiewicz, G., Zota, A.R., Fabian, M.P., Chahine, T., Julien, R., Spengler, J.D., & Levy, J.I. 2011. Moving environmental justice indoors: Understanding structural influences on residential exposure patterns in low-income communities. *American Journal of Public Health*, 101(S1), S238–S245.

Adams, J. 1995. *Risk*. London: UCL Press.

Adams, J. 2003. 'Risk and Morality: Three Framing Devices' in Ericson, R., & Doyle, A. (eds.). *Risk and Morality*. Toronto: University of Toronto Press. pp 84–103.

Aronson, N. 1984. 'Science as a Claims-Making Activity: Implications for Social Problems Research' in Schneider, J. & Kitsuse, J.I. (eds.). *Studies in the Sociology of Social Problems*. Norwood, NJ: Ablex. pp 1–30.

Blair, A., Saracci, R., Vineis, P., Cocco, P., Forastiere, F., Grandjean, P., . . . & Vainio, H. 2009. Epidemiology, public health, and the rhetoric of false positives. *Environmental Health Perspectives*, 117(12), 1809–1813.

Briggs, D.J., Sabel, C.E., & Lee, K. 2009. Uncertainty in epidemiology and health risk and impact assessment. *Environmental Geochemistry and Health*, 31(2), 189–203.

Brown, P. 1992. Popular epidemiology and toxic waste contamination: Lay and professional ways of knowing. *Journal of Health and Social Behavior*, 33, 267–281.

Carlsson, F., & Johansson-Stenman, O. 2000. Willingness to pay for improved air quality in Sweden. *Applied Economics*, 32(6), 661–669.

CDC. 2013. Investigating suspected cancer clusters and responding to community concerns. *Morbidity and Mortality Weekly Report, Recommendations and Reports*, 62(8), 1–14.

COEHHA. 2001. *Saccharin Delisted*. Available at: http://www.oehha.ca.gov/prop65/CRNR_notices/chemicals_reconsideration/fdelistsacc.html (Accessed 10 April 2015).

Costas, K., Knorr, R.S., & Condon, S.K. 2002. A case–control study of childhood leukemia in Woburn, Massachusetts: The relationship between leukemia incidence and exposure to public drinking water. *Science of the Total Environment*, 300(1), 23–35.

Dourson, M.L., Felter, S.P., & Robinson, D. 1996. Evolution of science-based uncertainty factors in noncancer risk assessment. *Regulatory Toxicology and Pharmacology*, 24(2), 108–120.

Eurostat. 2012. *EU Transport in Figures: Statistical Pocketbook 2012*. Available at: http://ec.europa.eu/transport/facts-fundings/statistics/doc/2012/pocketbook2012.pdf (Accessed 4 June 2015).

Fuchs, M. 1996. Woburn's burden of proof: Corporate social responsibility and public health. *Journal of Undergraduate Science*, 3, 165–170.

Gardner, M.J., Winter, P.D., & Acheson, E.D. 1982. Variations in cancer mortality among local authority areas in England and Wales: Relations with environmental factors and search for causes. *British Medical Journal Clinical Research*, 284(6318), 784–787.

Goodman, M., Naiman, J.S., Goodman, D., & LaKind, J.S. 2012. Cancer clusters in the USA: What do the last twenty years of state and federal investigations tell us? *Critical Reviews in Toxicology*, 42(6), 474–490.

Grandjean, P., Satoh, H., Murata, K., & Eto, K. 2010. Adverse effects of methylmercury: Environmental health research implications. *Environmental Health Perspectives*, 118, 1137–1145.

Grant, W.B. 2009. How strong is the evidence that solar ultraviolet B and vitamin D reduce the risk of cancer? An examination using Hill's criteria for causality: An examination using Hill's criteria for causality. *Dermato-Endocrinology*, 1(1), 17–24.

Hänninen, O., Knol, A.B., Jantunen, M., Lim, T.A., Conrad, A., Rappolder, M., . . . & EBoDE Working Group. 2014. Environmental burden of disease in Europe: Assessing nine risk factors in six countries. *Environmental Health Perspectives*, 122, 439–446.

Hansen, S.F., Krayer von Krauss, M.P., & Tickner, J.A. 2007. Categorizing mistaken false positives in regulation of human and environmental health. *Risk Analysis*, 27(1), 255–269.

Hansen, S., & Tickner, J. 2011. The precautionary principle and false alarms – lessons learned. In European Environment Agency, *Lessons Learned from Health Hazards*. Copenhagen: EEA. pp 17–45.

Harwood, V.J., Levine, A.D., Scott, T.M., Chivukula, V., Lukasik, J., Farrah, S.R., & Rose, J.B. 2005. Validity of the indicator organism paradigm for pathogen reduction in reclaimed water and public health protection. *Applied and Environmental Microbiology*, 71(6), 3163–3170.

Hill, Austin Bradford. 1965. The environment and disease: Association or causation? *Proceedings of the Royal Society of Medicine*, 58(5), 295–300.

Howick, J., Glasziou, P., & Aronson, J.K. 2009. The evolution of evidence hierarchies: What can Bradford Hill's 'guidelines for causation' contribute? *Journal of the Royal Society of Medicine*, 102(5), 186–194.

Hunter, P.R., & Fewtrell, L. 2001. 'Acceptable Risk' in Fewtrell, L., & Bartram, J. (eds.). *Water Quality: Guidelines, Standards and Health*. London: IWA Publishing. pp 207–227.

Jacobs, D.E., Clickner, R.P., Zhou, J.Y., Viet, S.M., Marker, D.A., Rogers, J.W., . . . & Friedman, W. 2002. The prevalence of lead-based paint hazards in US housing. *Environmental Health Perspectives*, 110(10), A599–A606.

Jamieson, D. 1996. Scientific uncertainty and the political process. *The Annals of the American Academy of Political and Social Science*, 545, 35–43.

Lemons, J., Shrader-Frechette, K., & Cranor, C. 1997. The precautionary principle: Scientific uncertainty and type I and type II errors. *Foundations of Science*, 2(2), 207–236.

Mayeux, R. 2004. Biomarkers: Potential uses and limitations. *NeuroRx: The Journal of the American Society for Experimental Neuro Therapeutics*, 1, 182–188.

McKenzie, L.M., Guo, R., Witter, R.Z., Savitz, D.A., Newman, L.S., & Adgate, J.L. 2014. Birth outcomes and maternal residential proximity to natural gas development in rural Colorado. *Environmental Health Perspectives*, 122(4), 412–417.

Morin, A., Brook, J.R., Duchaine, C., & Laprise, C. 2012. Association study of genes associated to asthma in a specific environment, in an asthma familial collection located in a rural area influenced by different industries. *International Journal of Environmental Research and Public Health*, 9(8), 2620–2635.

Morris, J.N. 1975. *Uses of Epidemiology*. Edinburgh: Livingston.

Murray, C.J., Vos, T., Lozano, R., Naghavi, M., Flaxman, A.D., Michaud, C., Ezzati, M., Shibuya, K., Salomon, J.A., Abdalla, S., Aboyans, V., Abraham, J., Ackerman, I., Aggarwal, R., Ahn, S.Y., Ali, M.K., AlMazroa, M.A., Alvarado, M., Anderson, H.R., & Anderson, L.M. 2012. Disability-adjusted life years (DALYs) for 291 diseases and injuries in 21 regions, 1990–2010: A systematic analysis for the Global Burden of Disease Study 2010. *The Lancet*, 380(9859), 2197–2223.

National Audit Office. 2009. *Improving Road Safety for Pedestrians and Cyclists in Great Britain*. London: NAO.

Nazir, M., & Khan, F.I. 2006. Human health risk modeling for various exposure routes of trihalomethanes (THMs) in potable water supply. *Environmental Modelling & Software*, 21(10), 1416–1429.

Needleman, H.L., & Gatsonis, C.A. 1990. Low-level lead exposure and the IQ of children: A meta-analysis of modern studies. *Journal of the American Medical Association*, 263(5), 673–678.

OECD. 2014. *The Costs of Air Pollution*. Geneva: OECD.

Ontario Medical Association. 2005. *The Illness Costs of Air Pollution: 2005–2026 Health and Economic Damage Estimates*. Toronto: OMA.

Petticrew, M.P., Sowden, A.J., Lister-Sharp, D., & Wright, K. 2000. False-negative results in screening programmes: Systematic review of impact and implications. *Health Technology Assessment*, 4(5), 1–120.

Priebe, P.M., & Kauffman, G.B. 1980. Making governmental policy under conditions of scientific uncertainty: A century of controversy about saccharin in congress and the laboratory. *Minerva*, 18(4), 556–574.

Pukkala, E., Martinsen, J.I., Lynge, E., Gunnarsdottir, H.K., Sparén, P., Tryggvadottir, L., . . . & Kjaerheim, K. 2009. Occupation and cancer-follow-up of 15 million people in five Nordic countries. *Acta Oncologica*, 48(5), 646–790.

Sakai, T. 2000. Biomarkers of lead exposure. *Industrial Health*, 38(2), 127–142.

SENES Consultants Limited. 2012. *Health Impacts Exposure to Outdoor Air Pollution in Hamilton, Ontario*. Hamilton: Clean Air Hamilton.

Strimbu, K., & Tavel, J.A. 2010. What are biomarkers? *Current Opinion in HIV and AIDS*, 5(6), 463.

Swaen, G.G., Teggeler, O., & van Amelsvoort, L.G. 2001. False positive outcomes and design characteristics in occupational cancer epidemiology studies. *International Journal of Epidemiology*, 30(5), 948–954.

Thacker, S.B., Stroup, D.F., Carande-Kulis, V., Marks, J.S., Roy, K., & Gerberding, J.L. 2006. Measuring the public's health. *Public Health Reports*, 121(1), 14–22.

Thirty Years After Dioxin Disaster, EPA Returns to Times Beach. 2012. *The New York Times*. Available at: http://blogs.riverfronttimes.com/dailyrft/2012/06/dioxin_times_beach_new_epa_tests.php (Accessed 11 April 2015).

Tong, S., Von Schirding, Y., & Prapamonto, T. 2000. Environmental lead exposure: A public health problem of global dimensions. *Bulletin of the World Health Organization*, 78, 1068–1077.

van Leeuwen, F.R., Feeley, M., Schrenk, D., Larsen, J.C., Farland, W., & Younes, M. 2000. Dioxins: WHO's tolerable daily intake (TDI) revisited. *Chemosphere*, 40(9), 1095–1101.

Von Glasersfeld, E. 1987. *The Logic of Scientific Fallibility*, Paper presented at the 8th Biennial Conference, Mental Research Institute, San Francisco. Available at: http://www.vonglasersfeld.com/113.2 (Accessed 13 April 2015).

Wang, H., & Mullahy, J. 2006. Willingness to pay for reducing fatal risk by improving air quality: A contingent valuation study in Chongqing, China. *Science of the Total Environment*, 367(1), 50–57.

Welshons, W.V., Nagel, S.C., & vom Saal, F.S. 2006. Large effects from small exposures: III. Endocrine mechanisms mediating effects of Bisphenol A at levels of human exposure. *Endocrinology*, 147(6), s56–s69.

WHO. 1993. *International Programme on Chemical Safety Biomarkers and Risk Assessment: Concepts and Principles*. Geneva: Author.

WHO. 2000. *Principles for Modelling Dose–Response for the Risk Assessment of Chemicals*. Copenhagen: Author.

WHO. 2013. *Global Status Report on Road Safety: Supporting a Decade of Action*. Geneva: Author.

WHO. 2014. *Global Health Observatory – Disability-Adjusted Life Years*. Available at: http://www.who.int/gho/mortality_burden_disease/daly_rates/text/en/index1.html (Accessed 20 April 2015).

Wildavsky, A. 1997. *But Is It True? A Citizen's Guide to Environmental Health and Safety Issues*. Boston: Harvard University Press.

Wilde, G. 1994. *Target Risk*. London: PDE Publications.

World Bank. 1998. *Phasing Out Gas from Gasoline*. Technical Paper 397. Washington, DC: Author.

4 Is it likely to happen?

Introduction

Why do people hesitate to drink from plastic bottles and will fight fiercely against a nuclear waste facility near their home, but they are still willing to drive a car or take a bath? By some calculations driving a car and bathing are far more hazardous and 'risky' activities. So why will people expend considerable resources to avoid plastic water bottles and nuclear waste facilities? One school of thought is that we are irrational, we are simply misinformed about the *real* threats from potential hazards, particularly rare ones like nuclear waste or bisphenol A in food containers. Another theory suggests that the degree to which we are *voluntarily exposed* is at the heart of such differences and related to this; people *dread* some hazards but not others. Voluntariness and dread are two key elements of some early explanations for this curious gap in the way we seem to think about and act towards various environmental hazards.

In Chapter 3 we learned about some historical case examples of how difficult it is to calculate technological hazards risk and how seemingly small hazards at the population level become serious concerns and policy conundrums at the local scale – in communities. Chapter 4 will focus more attention on studies relating to risk in society and will further elaborate some of the assumptions and theories that are used to explain how we manage environmental health risks within society. Two approaches to understanding risk are singled out for particular attention – the revealed preference and the expressed preference approaches – since they had an early and strong influence on how academics think about the ways society deals with risk. These approaches are set apart from the social theories of risk (e.g., risk society, cultural theory and social amplification of risk) discussed elsewhere in the book since the expressed preference approach (also called the psychometric paradigm) in particular has prompted many spin-off studies. These early academic studies were more pragmatic and policy oriented than the social theories of risk, in that the revealed and expressed preference approaches attempted to answer a simple, and now infamous question within the world of risk research, "How safe is safe enough?" Yet the approaches to answering this simple question raised a whole host of new questions that reinforced the need for more sophisticated theories to explain individual and societal risk choices. This chapter does address the issue of the acceptability of risk and what predicts that acceptability and does not detail how risks are calculated by environment and health agencies for specific substances – for example, the tolerable daily intake of methylmercury-exposed fish (see Chapter 2).

Revealed risk preferences

One of the best-known and early approaches to addressing the question "How safe is safe enough?" came from Chauncey Starr (1969). His landmark study was based on large amounts of data concerning routine activities in society that have some element of risk associated with them – now known as the revealed preference approach. There are at least seven assumptions underlying his analysis. First, when people engage in an activity – like flying in a plane, rock climbing or consuming energy from burning coal – they tacitly accept the risks involved. Tacit acceptance is inferred from the fact that people actually pay to engage in these activities or benefit from these technologies. In this sense, technological hazards are meaningfully separated into voluntary and involuntary activities whereby for example hunting or smoking are activities in which we willingly choose to engage, while we may not choose to be exposed to mining accidents or electrocution from working in the electricity industry; yet we tend to benefit from such facilities in terms of having a stable electricity supply. Starr highlights that these facilities and activities may also be differentiated from each other in terms of the amount of time it takes to switch to a reasonable substitute. We may stop smoking or driving more or less immediately, but changing the exposure to the involuntary activities by switching out those technologies takes much longer because they often depend on (sometimes daunting) social and institutional changes. This issue is more acute at the local level whereby, for example, a mining community or other similar community dependent on a local resource (e.g., coal or uranium) may not *easily* switch to an economy based on alternative, cleaner, safer, renewable resources (e.g., solar or wind electricity generation).

Second, as suggested in the first assumption, there is a relationship between the benefits people enjoy from an activity or technology and the potential costs of the activity. Third, the benefits can be measured in terms of how much we spend as a society on the activity. For example, the benefits of smoking are based on the amount the average smoker spends on cigarettes and other tobacco products annually, while the benefits of mining are the average amount of income created for employees of the mining sector. Fourth, the costs or 'risks' can be measured in terms of the number of deaths that result directly from an activity as a ratio of the time spent on that activity – the fatalities per person-hour of activity (see Figure 4.1). Fifth, the historical accident records data are an accurate representation of the fatalities related to a technology or activity. Sixth, we arrive at the mix of technologies we currently prefer based on societal 'trial and error' over time. Seventh, and despite the trial-and-error assumption, the ratio of benefits to fatalities will remain relatively constant over time. These assumptions lay the groundwork for an analysis whose purpose was to reveal some underlying 'law' of social behaviour regarding risk. Thus, Starr's cross-calculation of risk ratios (risks to benefits) at a single point in time was presumed to tap into a much more extensive and temporally invariant process of individual and societal risk decision making. Starr's analysis was not based on individuals, but instead he analyzed aggregate, society-level data. He added up all of the money society spends on each activity to represent the aggregate benefit to society for that activity; similarly all the fatalities for the activity represented the risk. We note the difference between individual and aggregate data here since it is a theme throughout the book, and we return to this idea in the discussion of the limitations of the revealed preference approach farther on in the chapter.

Key findings of the revealed preference approach

There were three main conclusions to Starr's study. The first is that, perhaps not surprisingly, there is a correlation between risk and benefit. In this case, the greater the benefit an activity provides for society, the greater the risk of harm that society is willing to accept. That is, the lines for risk are positively sloped in Figure 4.1, and Starr estimated that the acceptability of risk is proportional to the third power of the benefits. Second, his study suggests we are willing to accept voluntary risks that are 1,000 times greater than involuntary risks. As Figure 4.1 highlights the voluntary and involuntary risk lines of best fit have the same slope, but the voluntary line is shifted up by three decimal places. For example, Americans in the late 1960s would pay to accept the fatality risks of engaging in various sporting activities (e.g., skiing) that are roughly equivalent to the risk of dying from disease (about 2×10^6 fatalities per person-hour exposure for a less than $500 average annual cost to participate in that sport). By comparison while people were less likely to die from exposure to electricity (2×10^{-9} fatalities per person-hour exposure) they also derived more benefit (about $700 per person per year based on the contribution of electricity to the gross domestic product). The third main finding is that the risk of disease is a 'psychological yardstick' against which society seems to gauge all other risks (dotted line in Figure 4.1). Thus, those activities that involve a higher risk of fatalities will require proportionally higher benefits, while those with a lower risk of fatalities will require fewer benefits, keeping in mind the divide between involuntary and voluntary risks.

It may be useful, too, to put Starr's work in its historical context in the sense that this was a time of great potential for nuclear power as a 'cleaner and more efficient' alternative to the use of fossil fuels for electricity generation. Starr himself was an electrical engineer who worked in the nuclear industry. Thus, his efforts may be viewed as a rationale for showing society why nuclear power – when designed to a specific standard – should be acceptable, that is, since we engage in so many other activities that pose even greater risks of harm than nuclear power. There is a specific section near the end of the paper titled "Atomic Power Plant Safety" in which Starr suggest that nuclear plants might be socially acceptable at 1/200th or 1/40th the risk level (i.e., higher safety) of coal-powered electricity. This is based on the assumption of one 'catastrophic failure' (about 10 deaths)

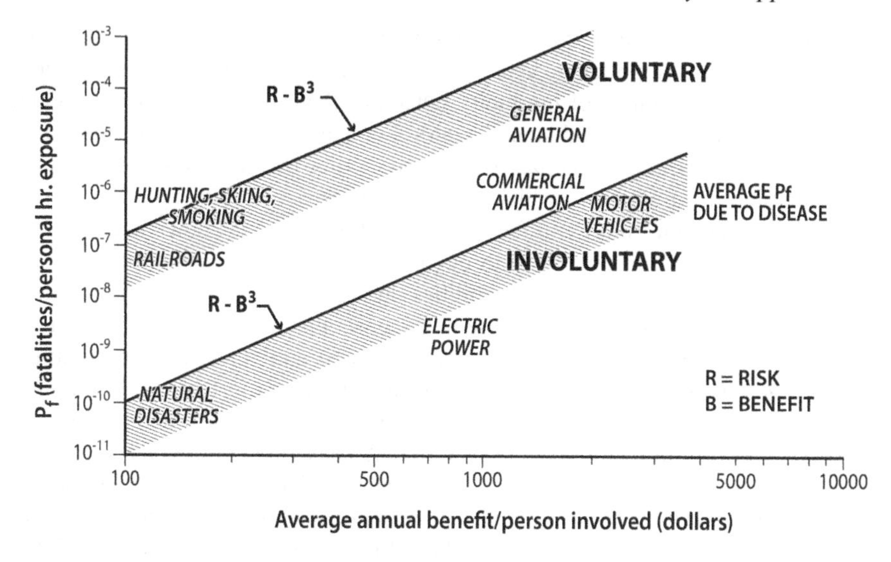

Figure 4.1 Revealed risk preferences

Source: Starr (1969), reprinted with the kind permission of AAAS.

in 100 plant years of nuclear energy operation. In fact Starr (1969, p. 1237) is quoted in the paper as follows: "*Can a nuclear power plant be engineered with a predicted performance of less than 1 catastrophic failure in 100 plant years of operation? I believe the answer is yes*". This is particularly interesting given that the 1960s was still a time when scientists, even social scientists, refrained from stating their own opinions, particularly those that were well beyond the data they were presenting. Thus, this statement can be viewed as a political statement, an issue to which we will return throughout this book. Yet there are a number of other limitations to this type of analysis and thinking, and it is to these that we now turn.

Limitations of the revealed preference approach

There are a number of limitations to Starr's work, many of which he himself points out in his (1969) article in *Science* and subsequent writing; and these limitations have been an important springboard for follow-up research. Many of these limitations are of direct relevance to the core theme of this book: the social dimensions of environment and health relationships. Five such limitations are discussed in what follows: (i) fatalities are not the only measure of harm; (ii) relatively new or rare hazards are difficult to assess; (iii) societies do not always have choice in the form of reasonable substitutes; (iv) the potential ambiguity of the term 'voluntary'; and (v) the implied irrationality of alternative, conflicting assessments of technological hazard risk. Of course all science requires simplifying assumptions to make the calculations possible (see Chapter 3), but it seems that *too* much is assumed by Starr which potentially renders equally important issues invisible along the way.

First, if risk is defined as the probability of harm along with the type and scale of that harm, it is important to keep in mind that there other forms of harm besides death. There is a range of morbidity outcomes that are not considered including impacts to physical and mental health. For example, air pollution may cause or exacerbate various pulmonary (lung) and cardiac (heart) conditions including asthma and heart attacks – which may in turn restrict activities of daily living. Energy production from fossil fuels and coal in particular is known to exacerbate air pollution problems directly or through the contribution to ground-level ozone. Starr's response to this is that disabilities due to illness are not readily reduced to a common currency. In this sense he suggests that at the time of his analysis introducing morbidity into the calculations would add "inconvenient complexity" (1969, p. 1234). Yet our definition of health set out in Chapter 1 combines morbidity and mortality to even encompass issues of individual and social well-being. The disability-adjusted life year (DALY) described in Chapter 3 – a measure that considers whether disease, illness and particularly disability are part of daily living (WHO, 2002) – would be an example of a broader and more inclusive measure of negative impacts of technology. Since Starr's (1969) article, much has been studied and written about the social impacts of technological hazards – for example, the psychosocial impacts on local communities from facilities including landfills, incinerators and nuclear power plants (Beardslee and Mack, 1982; Elliott et al., 1993; Eyles et al., 1993). Psychosocial impacts to local residents can come from uncertainty about other health effects (e.g., cancer, asthma), and efforts to fight against facilities can have powerful implications for individual and family stress (Baum, 1993) as well as localized social conflict over potential facility harms (Edelstein, 2004). It is very difficult too to predict the social impacts of catastrophe in terms of displacement (evacuation), livelihood losses and changes to ways of life, particularly for new hazards and technologies. For example, Starr likely underestimated the range of human, social and environmental impacts of a nuclear accident on the scale of Chernobyl[1] or, more recently, Fukushima.

Second, newer technological hazards – ones often found high on public policy agendas – are particularly difficult to assess. Regardless of the outcomes, there are typically too few events to do an assessment of harm with a sufficient level of confidence. Starr himself points out that data on fatalities due to environmental pollution are particularly problematic since such linkages are likely multi-causal and complex; that is, they are difficult to reduce to direct cause and effect (see Chapters 1 and 3). In this light, the calculation of harm from vehicles is too over-simplified if reduced to the fatalities from accidents since, for example, the emissions from those vehicles and the production of the fuel both may contribute to several negative health and well-being impacts. Yet for vehicles we at least have the luxury of several years of data on emissions to do those complex assessments if we so choose. This is not the case for newer hazards for which mortality and morbidity events are, by definition, lacking. In this sense society is faced with policy conundrums such as, "Will substance X or technology Y ultimately cause thousands of cases of cancer that may or may not lead to death?" In some cases a precautionary approach is taken which results in a ban or severe restrictions on the use of the substance in certain places – for example, DDT pesticide, urea formaldehyde foam insulation (UFFI), and thalidomide in Canada. Ironically, as with DDT, UFFI, and thalidomide this is often done after the substance has been in use for several years and potential harms are detected through post hoc population health (epidemiology) and laboratory (toxicology) studies. It is very difficult to know when to take prophylactic action because there is considerable uncertainty involved in defining and predicting harms ahead of approving the use of such substances. As discussed in Chapter 10 though, we do not always seem to learn from our mistakes.

For many modern-day environmental hazards risk is simply not calculable according to Starr's method. This is particularly salient because at the time Starr wrote the paper there was a subtext of trying to understand aversion to nuclear power plants. One of the practical problems, which Starr himself recognized, is that it takes several years, often decades, to determine the error and hence fatalities that may result from a complex technology like nuclear power generation. Quite simply, the jury is out on the number of accidents that would result in fatalities – the number is unknown and in some cases unknowable. Further, in the case of nuclear power and other modern-day hazard concerns, the negative impacts and premature death are difficult to track since outcomes like cancer may take several years to manifest and are multi-causal. In 1969 when Starr published the paper, few if any nuclear power plant deaths had occurred particularly from radiation, and he suggested that it would take at least 30 years to have a reasonable assessment of harms. However, since then there have been several high-profile nuclear accidents including Three Mile Island (1979), Chernobyl (1986) and most recently Fukushima (2011). Though Chernobyl in particular can be linked to several premature deaths, such deaths likely number in the thousands[2] not the tens or hundreds of thousands that would be required to elevate the risk of nuclear power plant fatalities beyond the 10^{-9} that Starr assessed several decades ago when nuclear power had only been operational in the U.S. for about 10 years. Equally troubling is the fact that current medical and social scientific methods are only modestly equipped to understand the impacts of technologies like nuclear power. For instance, the main report from the World Health Organization on the impacts of Chernobyl laments that distinguishing facility radiation leak–induced cancers from normal variation in cancer is, "very difficult with available epidemiologic tools" (Chernobyl Forum, 2006, p. 16). This general problem of a lack of data on mishaps for new technologies and the unknowability of impacts may explain why there have not really been any attempts to update and replicate Starr's method with more modern-day hazards. Actually, there are more nuanced explanations for why people want to continue to avoid involuntary risks like nuclear plants, despite the low probability of disaster. Whitfield et al. (2009) provide such an example in their study of public risk perception of nuclear power in the U.S. More than two decades after the Chernobyl and Three Mile Island disasters, they find that widespread opposition to nuclear power had given way to public 'ambivalence' about using this technology. Their findings may partially explain a surge in the nuclear industry leading up to the 2011 Fukushima disaster. Such explanations align with the revealed preference approach which is discussed in the next section, as well as other cultural (e.g., cultural theory of risk) and social (e.g., social amplification of risk) frameworks.

The third limitation of Starr's approach is the assumption that society arrives at its current mix of technology use by trial and error since it is tenuously based on the idea that there are reasonable substitutes for all technologies and activities. That is, there is a supposition that we are making conscious risk/benefit trade-offs at both the level of the individual and of society. However the substitutes are often not available either literally or at a cost that is reasonable for most in society. For example, drivers may desire to substitute their gasoline vehicle with a less polluting biodiesel, ethanol or electricity version but are deterred by both vehicle availability (including costs) and structural barriers to the local availability of fuel sources. These choices are also spatially constrained in the sense that some technologies are particularly poorly developed locally. These problems are magnified for new activities and technologies that may take decades to get a foothold. Alternatives to this assumption also include the *risk compensation* thesis, that once lower risk is achieved through regulation, people behave in such a way as to bring the risk of that activity back up to where it was before the regulation. Adams and Hillman (2001) for example demonstrate how cyclists take greater risks when riding with a helmet to negate the protective effect of wearing a helmet. Returning to the issue of alternative fuels sources, we may simply consume the same amount of energy over time as vehicle efficiency increases the number of vehicles, or distances driven per capita may increase may likewise increase – a phenomenon similar to risk compensation called the *rebound effect* (Small and Van Dender, 2007). Starr's assumption that society tries out a technology, learns the real costs and benefits and then readily moves on to the next, better technology in a seamless, socially even and barrier-free way is thus fraught with difficulties.

Fourth the revealed preference approach is reductionist as the decision-making processes themselves are left invisible. How do we arrive at the current mix of technologies in widespread use? What do people think about the decisions we have made as a society? How does society handle groups who prefer not to be exposed to particular hazards perhaps because they are vulnerable? Some of this is captured in Starr's distinction between voluntary and involuntary risks, but such a distinction only hints at the range of social, political and cultural aspects of technological hazard risk. Overall the approach leaves the impression that any social value that individuals or groups hold about an activity that deviates from the lines of best fit – for example, those who are more precautionary – are at worst deviant and irrational. That is, if people are willing to drive a car or fly in a plane, why are they not willing to accept current safety standards for nuclear power generation? That many want or choose to avoid the latter technology but not the former two attests to the idea that there is much more involved – the social, the cultural and the political.

Despite these limitations Starr's approach has had a powerful influence in the world of risk research and, by extension, social scientific approaches to environment and health. Perhaps the most influential aspect of his publication in *Science* and subsequent similar work was to engage academics and other publics in discussion and debates about how to understand risks from technology and their impacts on environment and health. One of the products of these debates is the expressed preferences approach to understanding risk.

Expressed risk preferences

The expressed preference approach directly addresses some of the key limitations of the revealed preference approach, specifically in terms of understanding what people actually think of various technologies. There are at least two early branches of this approach, the first based largely in experimental decisions under uncertainty using hypothetical gambles (e.g., Kahneman and Tversky, 1984) and the other, very much connected to the first, concerns attitudes towards specific environment and health hazards and the degree to which people feel such activities need tighter regulation (e.g., Slovic, 1987).

We are loss averse

In Kahneman and Tversky's experimental studies participants were asked to choose one of two gambles:

A – 85% chance to win $1,000 (15% chance of $0)
B – 100% chance to win $800

Most people pick B even though according to rational economic choice people would be expected to pick A, since it has a slightly better utility. That is, 0.85 × $1,000 + 0.15 × $0 = $850, which is $50 greater than 1.0 × $800 = $800. This represents one of the key features of the psychology of gambles – we are averse to losses. This is one of the main conclusions made by Bernoulli, who defined the relationship between gains and losses according the hypothetical curve in Figure 4.2

The key feature of this asymmetrical curve is that the slope of the line for losses is steeper such that a loss of $500 is (dis)valued more than a gain of $500 is valued. What this loss aversion does though is create a situation in which people are prone to take gambles involving losses and not take gambles involving gains. To demonstrate this principle further Kahneman and Tversky experimented with a choice between the following two gambles:

C – 85% chance of losing $1,000 (15% of losing $0)
D – 100% chance of losing $800

Most people pick C, as the majority of people prefer to take a gamble if there is a chance they will prevail by losing nothing despite only a meagre 15% chance. Yet, as with the first example above involving gains, summing the gambles shows that D would be the rational choice with a calculated value of only −$800 compared to C's −$850 (0.85 × −$1,000 + 0.15 × $0). This loss aversion, particularly

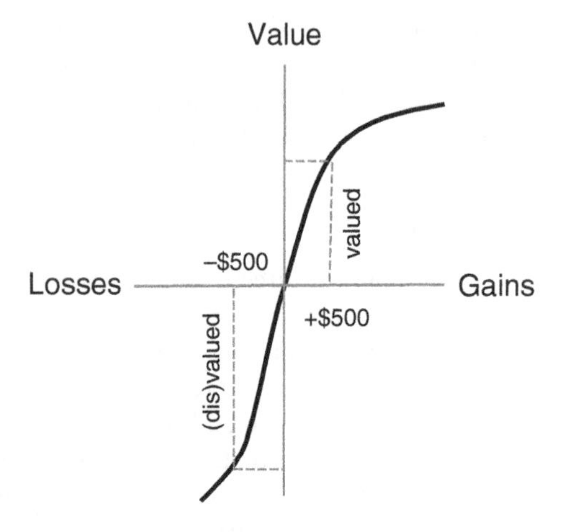

Figure 4.2 Hypothetical risk value function

Source: Adapted from Kahneman and Tversky (1984).

the gamble between C and D, is the basis upon which the insurance industry flourishes since people are willing to spend substantial sums of money to prevent events that are uncertain yet improbable.

Kahneman and Tversky along with various psychologists extended this basic idea to show that we generally do not make choices according to the rational calculus of summing the probability multiplied by gains and losses. Instead they suggest we are expected to make decisions based on how choices are framed. Above all we react differently depending on whether a gamble is framed as a gain (the choice between A and B – most pick B) compared to when it is framed as a loss (the choice between C and D – most pick C).

One of the main implications of this work for understanding environmental and health is that people will not react according to the utility calculus that experts might use. The second implication is that how a choice is framed can influence how we view decision paths relating to technological hazards. This further suggests that communication about environment and health risks and other social factors may be as important as the risks themselves.

Heuristics and biases

Heuristics and biases are two concepts used to describe decision making under uncertainty – an area of research most commonly linked to cognitive psychology. The results of the expressed and revealed preference studies are sometimes summarized as heuristics and biases of risk perception. A heuristic is a simplifying assumption or mental shortcut we use to make a decision. Examples of common heuristics are the *availability heuristic* and the *anchoring and adjustment heuristic*. The availability heuristic is based on the idea that we tend to judge an event according to the ease with which we can bring instances of that event to mind.

This 'ease' tends to have both temporal and spatial aspects in the sense that the recency of our exposure to such a negative event from a facility/technology (e.g., a leak) and the spatial proximity of the event to the individual tend to make the event easier to access in our memory. Thus, people may be more inclined to assign a high probability to future industrial accidents, hurricanes or even nuclear accidents if one has happened recently and/or nearby. The availability heuristic may be invoked to understand the role of the media in risk perception such that if we can easily bring an event to mind we tend to assess it as more probable.

The anchoring and adjustment heuristic is typically demonstrated in terms of the use of numbers as anchors, or first impressions, which we tend to use to make choices. Tversky and Kahneman (1974) used a series of simple experiments to demonstrate this idea – for example, they divided a group of individuals in half and assigned the individuals in one group to estimate the product of E and the other group to estimate the product of F within 5 seconds, and without any aids:

E: $1 \times 2 \times 3 \times 4 \times 5 \times 6 \times 7 \times 8$
F: $8 \times 7 \times 6 \times 5 \times 4 \times 3 \times 2 \times 1$

The median estimate for the E group tends to be much lower than for F (in the published experiment the results were 512 and 2,250 respectively, while the actual answer is 40,320). Since we are apt to read from left to right, it is reasonable to assume that the first number is the anchor. Since E starts with the anchor of 1 or perhaps even 1×2 while F has the anchor of 8 or perhaps 8×7, the final estimate is based on a rough guesstimate once a person completes the calculations that can be made in the first 5 seconds.

Thus, if we are given a number before being asked to make a decision about some subsequent preference, we tend to adjust according to that initial number. This can be related to environment and health in a number of ways. For example, if we are made aware of a set of mortality probabilities (e.g., from driving an automobile for a year) before making a decision about the mortality probabilities for a new technological hazard (e.g., a waste incinerator technology) there is a good chance our estimate will be based on how we feel the new hazard relates to the one we are given – that is, we will anchor to the automobile data and adjust according to that anchor. Tversky and Kahneman (1974) showed that when two groups were asked to estimate the percentage of African nations in the United Nations the group which observed a rigged roulette ball land on 10 just prior gave an average estimate of 25% while the group that observed the ball land on 65 gave an average estimate of 45%.

Reassessing the actual (expert) versus perceived (lay public's) risk dichotomy

One way the problem of risk has been framed in policy circles is to suggest that laypeople need to think more like experts, that is, in terms of the actual fatalities caused by various activities. On the contrary the researchers, who developed the expressed preferences idea, also found that laypeople are actually quite good at assessing the fatalities from various hazards. That is, Fischhoff et al. (1982) found that laypeople could put the relative order of deaths from various diseases and other hazards in a similar sequence to what the mortality statistics show. There was a relatively high correlation between the lay assessments of the number fatalities caused by 40 causes of death and recorded data for those causes. Nevertheless, while laypeople tend to underestimate various disease fatalities, they tend to overestimate so-called natural hazards like tornadoes and floods. Despite heuristics and biases, laypeople can assess fatalities relatively accurately, yet they also tend to want *more* safety for most technologies. This suggests that there is something about hazards besides fatalities that elicits the desire for greater safety regulation. This conundrum that laypeople know how many fatalities are involved yet still may desire higher regulation to reduce risk is the basis for the second approach to expressed risk preferences. Before turning to this second approach we explore how experts perceive the calculation of 'actual' risk under uncertainty for relatively new technologies – that is, when data are lacking.

The idea that there is an 'actual' risk that can be known to experts if the right methods are applied is suspect because the experts themselves can disagree. This issue is discussed more fully in Chapter 3 concerning the limitations of toxicological and epidemiological methods. Kraus et al. (1992) conducted a survey of professional toxicologists (the Society of Toxicology) who were compared to a sample of the public from Portland, Oregon. The authors expected that the lay public would diverge from expert toxicologists on all of the measures which included items such as the ones in Table 4.1 and that different types of toxicologists (industry, academic, regulatory) would have different views. Their hypotheses were generally supported, and what was perhaps most eye opening was that on several of the survey measures the expert toxicologists disagreed with each other considerably. In some cases this was simply a matter of slight degree whereby for example 59% "agreed" that "Use of chemicals has improved our health more than it has harmed it" while 33% "strongly agreed" with this statement – yet overall 92% agreed in one way or another with the statement. Many of the items they queried in relation to the meaning of dose-response relationships and the regulation of chemical risks differed mainly in this fashion with a clear majority either agreeing or disagreeing depending on the statement with the majority of the division between the 'somewhats' and the 'stronglys' and not across the agree–disagree divide.

What surprised the authors is that on several items the experts straddled the disagree–agree divide. Table 4.1 has several examples of this with, for example, 50% of the experts agreed that "The way that an animal reacts to a chemical is a reliable predictor of how a human would react to the same chemical," while 39% disagreed (the remaining 11% either strongly disagreed – 2%, strongly agreed – 5%, or had no opinion – 4%). In fact, looking down the Disagree and Agree columns in Table 4.1 shows that all but one of the items (2b) shows substantially conflicting opinion on "Trust in animal studies". This same disagreement is pervasive in the study with these experts disagreeing on several items related to

Table 4.1 Responses of toxicologists and laypersons to questions about trust in animal studies

	Questions		*Strongly disagree*	*Disagree*	*Agree*	*Strongly agree*	*Don't know/ no opinion*
2a	The way that an animal reacts to a chemical is a reliable predictor of how a human would react to the same chemical.	T	1.9	38.9	50.3	5.1	3.8
		P	5.5	40.2	40.2	3.5	10.6
2b	Laboratory studies of a chemical's harmful effects on animals will, if properly done, identify all possible harmful effects of that chemical.	T	26.8	56.1	15.2	1.2	0.6
		P	17.1	60.7	11.3	2.3	8.6
2c	Laboratory studies of a chemical's harmful effects on animals allow scientists to accurately determine how much of the chemical it takes to cause similar harm in humans.	T	12.2	55.5	29.3	1.2	1.8
		P	11.7	50.0	28.9	0.8	8.6
2d	If a scientific study produces evidence that a chemical causes cancer in animals, then we can be reasonably sure that the chemical will cause cancer in humans.	T	10.3	47.3	39.4	1.2	1.8
		P	1.9	22.9	64.0	5.4	5.8

T = toxicologists

P = public

Source: Kraus et al. (1992), reproduced with kind permission from Wiley.

general attitudes towards chemicals – fore example, 47% (36% strongly) disagree and 46% (38% strongly) agree that “Our society has perceived only the tip of the iceberg with regard to the risks associated with chemicals.” One explanation for such differences may be gender – the women in the study overall tended to be more risk averse to chemicals and suspicious of what we can know about them through toxicology and dose-response relationships. Another explanation is the types of toxicologists. When toxicologists were subdivided into three categories – academic, industrial and regulatory – there were some significant differences that may go some way to explaining some divergence among toxicologists’ views. For example, for the statement “There is no safe level of exposure to a cancer-causing agent,” 54% of academic toxicologists disagreed (18% strongly) and 70% of regulatory toxicologists disagreed (18% strongly), but a much larger 88% of industrial toxicologists disagreed (46% strongly). This pattern of disagreement is underscored in Table 4.1, which shows wide disagreement on the degree to which humans are considered more or less vulnerable to the effects of chemicals than the animals on which toxicological studies are conducted.

We prefer to control that which we dread (or how we die matters)

The second approach, like the first, has its roots in psychology but focuses more on specific human activities and substances. This is often called the psychometric approach because it involves the quantification of how people mentally assess various potential hazards. The approach also links to a classic problem that geographers had been wrestling with for several years prior to the 1980s when the psychometric approach gained considerable traction – Why do people move back into communities devastated by flooding since these areas are also prone to future catastrophic flooding? That is, why do individuals and communities apparently downplay high-probability, high-consequence hazards, yet they devote more attention and concern to low-probability (and possibly) relatively low-consequence hazards?

The psychometric expressed preference approach involves asking people how they assess various technologies according to a number of dimensions of those hazards including ‘voluntariness’ but also a host of others like whether we feel the risks are well known to science, whether the hazard has catastrophic potential or whether it is ‘dreaded’ (makes us feel uncomfortable at the level of a gut reaction). Thus, while the revealed preference approach was imputed from data at the level of society as a whole, the expressed preference approach has focussed on generalizing from the study of small samples of individuals and their perceptions. For example, a landmark expressed preference study by Fischhoff et al. (1978) tests Starr’s ideas about voluntariness by asking people directly. They studied 76 people (52 members of the League of Women Voters and 25 of their male spouses) by asking them to rate 30 different activities (8 of them the same as Starr’s) according to nine risk dimensions (voluntariness, immediacy of effect, knowledge about risk to those exposed, knowledge about risk to science in general, control, newness, catastrophic potential, dread and severity as certainty of fatality) as well as three other dimensions (perceived benefit, perceived risk and acceptability of current levels of risk).

They did find a relationship between risk and benefit, but in contrast to Starr's revealed preference approach which found a *positive* relationship between revealed risk and revealed benefit, Fischhoff et al. found a negative relationship between perceived risk and perceived benefit (Figure 4.3). Even the main outlier, motor vehicles, which had strong positive perceived benefits and risks, the risks being rated 247 on a 300-point scale while the benefits only rated 187. High-risk, low-benefit technologies included nuclear power, handguns, smoking, motorcycles and alcoholic beverages, while low-risk, high-benefit technologies included non-nuclear electric power, railroads, prescription anti-biotics and vaccinations. Though there was an apparent flip-flop in findings (slope of the line of best fit) between revealed to expressed preference approach when Fischhoff et al. graphed perceived benefit against the perceived acceptability of each activity's risk,[3] the line of best fit in that case was again in the positive direction. Yet most activities were perceived to require increased safety measures.[4] In terms of Starr's finding that voluntary risks are more acceptable than involuntary risks, Fishchoff et al. produced contrary results using the expressed preference method. Whereas Starr had a voluntary risk line shifted upwards three decimal places (Figure 4.1), voluntary and involuntary activities had very similar negative slopes (Figure 4.3) with lines of best fit that cross in the expressed preference iteration. However, when graphed against risk *acceptability* the results are similar to Starr's in that there is clear separation between the voluntary and involuntary risks. Thus, how risk and benefit are framed and measured (e.g., perceived risk as fatalities vs. perceived risk *acceptability*) can have a potentially profound effect on the results and potentially on action in terms of mitigation efforts.

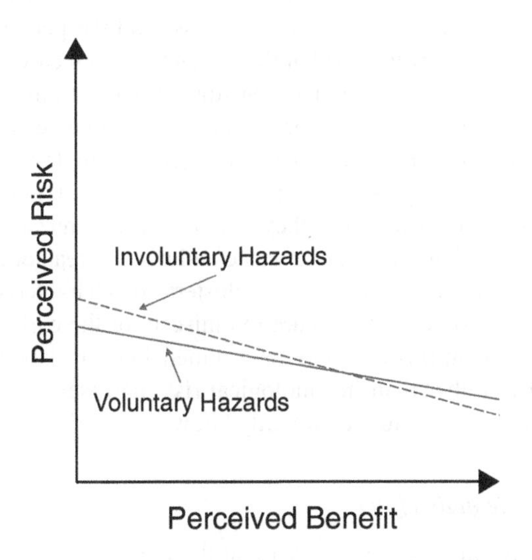

Figure 4.3 Expressed preferences for involuntary and voluntary hazards

Source: Adapted from Fischhoff et al. (1978).

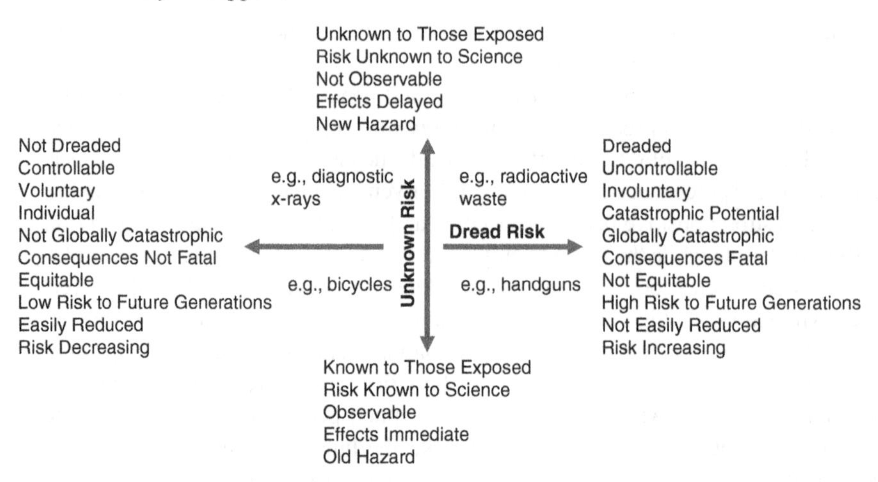

Figure 4.4 Clustering of 15 risk characteristics into dread and unknown risk

Source: Adapted from Slovic et al. (1980).

An influential aspect of this landmark study was the analysis of the nine so-called risk characteristics. As the characteristics were highly correlated with each other they were analyzed using factor analysis to determine how they cluster. This analysis distilled the variables to two key dimensions or factors which they labelled 'technological risk' (newness, voluntariness, known to science, hazards are known to those exposed, delay of consequences, and catastrophic potential) and 'severity' (certainty of fatalities, dread, catastrophic potential). As expected each of these was highly correlated with the acceptability of risk, and the perceived risk measures together accounted for over 50% of the variance in each case.

The factor analysis was replicated later on student populations by Slovic et al. (1980), who expanded both the number of activities/substances assessed (from 30 to 90) and the number of risk characteristics (from 9 to 18) to include such characteristics as affects me personally, many people exposed, risks and benefits inequitable and risks increasing; and they subdivided key hazards like nuclear power into uranium mining, radioactive waste and nuclear reactor accidents (see also Slovic, 1987). The two main factors (clusters) of characteristics and how they sort the 90 technological hazards are reminiscent of the earlier study and is depicted in Figure 4.4. In this case the main dimensions are labelled 'unknown risk' and 'dread risk' rather than 'technological risk' and 'severity' respectively, with the former labels being more commonly known in the risk literature.

Gender and the white male effect

Risk researchers have also tested the idea that the type of person perceiving the hazard influences whether they see it as too risky. One of the only consistent findings in terms of socio-demographic variables is gender. Several studies have shown that men tend to be less concerned about technological hazards than do women. This finding holds for the general as well as for specific populations like scientists, where the perceptions of women and men were compared using an expressed preference approach. There is evidence to suggest that this effect is likely less about biology than it is about sociology since once race or ethnicity is added into the equation, it is white men that tend to sift out as being least concerned about hazards of all types. Of course not all white men are unconcerned about hazards, but there is a minority of about 30% who tend to rate all hazards in these surveys as 'very low'. That is, this 30% skew the data so much that they tip the balance to make all males appear to have low concern about technological hazards (Flynn et al., 1994). Several phenomena may contribute to this effect including the fact that this 30% tend to trust experts more and hold anti-egalitarian worldviews (see Chapter 5), and these particular men are likely in positions of power where they have control over decisions involving technological hazards. This may reduce unknown, dread and control aspects of any technological hazard for this group. It is not difficult to imagine how those who may be involved in the management of technological hazards may be prone to trust any expert – somebody like them – who controls or regulates a hazard in some way.

Trust and stigma

Expressed preference risk perception research also expanded to consider social, cultural and political factors involved in the perception of technological hazards. It is through this work that trust, stigma and justice all emerged as important predictors of high concern about risk. These considerations move us even further from considering aspects that are attributed to the hazard and more towards aspects of the social, cultural and political systems that surround hazard risk. For example, whether a survey participant trusts the people who manage and are responsible for ensuring the safety of a technology or activity has emerged consistently as a predictor of concern and desire to further regulate various technological hazards. In this sense trust has to do with various ongoing social relationships – for example, between the perceivers and government, industry and agencies responsible for ensuring or reporting on safety. In some cases trust is relatively easy to sustain. It is not too difficult to imagine an industry that employs hundreds of local residents retaining the confidence of the majority in the surrounding community due to ongoing relationships. For example, Fried and Eyles (2011) trace contours of optimism and progressivism about hosting a low-level nuclear waste facility in a town currently employing hundreds at the local nuclear plant in Kincardine, Ontariom Canada. It turns out that there is more to this relationship than direct or even indirect benefits, including attention to safe facility management, something that is aptly captured in the general concept of trust. Contrasting Kincardine, Fried and Eyles recount the legacy of nuclear contamination from poor nuclear radiation management that began from the time when the industry was in its infancy in Port Hope, Ontario, Canada. More than just a lack of current employment from the industry evokes themes oriented away from trust including death and crime. Further, if a technology or facility is unfamiliar to a community it may take considerable effort on the part of industry and regulators to convince locals of the safety of the technology involved. Numerous studies have documented, for example, the animosity towards the high-level nuclear waste facility in Yucca Mountain, Nevada (Kunreuther and Easterling, 1990; Slovic et al., 1991a; Slovic et al., 1991b; Flynn et al., 1992). For example, Flynn et al. (1992) find that despite the offer of a range of sizeable state-wide economic benefits, such incentives do not statistically predict support from Nevada residents for the facility, while a lack of trust in the nuclear industry remains a strong predictor of *lack* of support for the facility.

The findings concerning trust have been particularly noteworthy to those who had high hopes for risk communication. That is, there was early optimism that by communicating effectively with people about technological hazards, lay views could be made to come in line with the way many experts think. On the contrary, trust is now generally seen as a key barrier to risk communication ever being effective on its own. This is due to what has become known as the 'asymmetry principle' – trust is easier to destroy than it is to create (Slovic, 1993). Thus, if a firm wants to build a new facility in a community it will take time to build trust that is sufficient to get local buy-in, but if accidents at the facility undermine that trust it might never return. Building trust in the first place can be doubly difficult if the technology involved is socially stigmatized (Slovic et al., 1991b). For example, nuclear power has most certainly been stigmatized (further) in the wake of the Chernobyl and Fukushima nuclear plant meltdowns, such that several jurisdictions throughout the globe are considering devolving nuclear power generation in favour of alternative sources with less catastrophic potential (Schneider et al., 2011; Wittneben, 2012).

The examples used thus far also suggest that attention needs to be paid to the *places* where technological hazards are situated. For example, thinking about a *hypothetical* nuclear power facility or nuclear waste repository may evoke different reactions than a nuclear power facility or nuclear waste repository proposed for local community X. There are many aspects of the local community that could serve to heighten or minimize trust, stigma and overall concern. These issues are dealt with more fully in Chapter 9, which considers both risk and place including the social amplification and attenuation of risk framework (SARF), but the following section outlines risk perception differences by country.

International social group differences

The nuclear power industry examples suggest that there is a temporal element to risk as new events (disasters) are experienced and absorbed by society, but there are also risk perception differences between social groups including the country where participants complete psychometric risk surveys. For example, Rohrmann (1994) found that there are significant differences in the way that Australians, New Zealanders and Germans view the individual acceptability of risks from potentially hazardous technologies (IA), societal benefits of those technologies (SB) and the magnitude of perceived adverse impacts (RM). Overall, Germans appeared to be least concerned about the magnitudes of risks from such hazards as nuclear power and asbestos production. Yet coal power plants were seen by German respondents to have a higher risk magnitude than they were viewed by their Australian and New Zealand counterparts (Table 4.2). By contrast, Australian and

Table 4.2 International risk perception comparisons: Australia, New Zealand and Germany

AUS/NZ/FRG (N = *129/130/80*)										
		RM			*SB*			*IA*		
Country:		*FRG*	*AUS*	*NZ*	*FRG*	*AUS*	*NZ*	*FRG*	*AUS*	*NZ*
B	Car racing	7.5	6.7	6.4	1.4	2.8	2.5	6.4	7.3	7.4
D	Asbest. prod.	8.0	8.4	8.2	4.2	3.2	3.4	3.4	3.1	3.6
G	Smoking	8.0	9.0	8.9	0.9	0.9	0.7	3.7	5.1	4.7
K	Firefighter	4.5	6.3	6.3	8.7	8.5	8.7	7.7	6.4	6.5
N	Coal p. plant	5.6	5.1	4.3	7.2	6.6	6.1	4.5	4.5	4.8
R	Earthquake ar.	6.0	6.7	5.8				5.1	5.4	5.9
U	Nucl. p. plant	6.1	7.0	7.0	5.0	4.9	4.4	4.0	3.4	3.3
W	Poll. urb. area	6.3	6.7	7.0				3.7	4.1	3.9
M	*(Mean)*	5.8	6.3	6.1	4.7	4.8	4.8	5.2	5.3	5.3

Source: Rohrmann (1994), reprinted with kind permission from Springer Science+Business Media.

New Zealand respondents rated the overall magnitude of the risks from and individual acceptability of a nuclear power plant to be lower. Thus, while there may not be overall differences (column totals), individual technologies and substances may have particular salience in certain contexts (individual rows). For example, firefighting is rated by Germans to have a relatively low risk magnitude combined with a relatively high acceptability level, which suggests something about either the safety of firefighting in Germany or how this occupation is socialized in Germany compared to the other two countries.

Conclusion

Are we overreacting to certain hazards like nuclear electricity generation because of innate heuristics and biases that we use to survive in most other social situations? Or is there something about risks from technology that laypeople simply need to better understand? Certainly the various studies cited in the chapter may give us pause about our abilities to accurately estimate risks under certain conditions, but heuristics and biases tell only part of the story of environment and health risks. This chapter describes how environmental hazard risk may be quantified at the population level by studying aggregate-level population data (revealed preference approach) and individual-level data on public perceptions (expressed preference approach). Both of these approaches address the question "How safe is safe enough?" for a society. What these approaches do not tell us in much detail is how risk is calculated and managed for specific technologies and substances. That is, how do experts actually calculate the risk and hence the safe exposure to a substance or technology? That is the topic of Chapter 3.

The current chapter also stops short of more sophisticated social *theorizing* about why we are concerned about some hazards and not about others. Though people are quite good at estimating the fatalities from particular activities and technologies, they may not necessarily use those estimations as the primary nexus for making decisions about how risky they feel an activity/technology is or whether it requires further regulation to make it acceptably safe. There are some important concepts and variables that provide important clues to what more comprehensive social theories of risk might look like, including voluntariness, dread, the role of benefits, trust, stigma, gender and place. The ideas are further developed in Chapter 5, which addresses various social theories of risk including the risk society, cultural theory of risk and the social amplification and attenuation of risk, along with a more detailed treatment of the promise and limitations of risk communication as a mechanism for managing risk.

Notes

1 World Health Organization estimates are that there were approximately 9,000 excess (thyroid) cancers, but more than half of those are in children under the age of 18; 350,000 evacuees; traumatic social dislocation and widespread psychological trauma. The human health toll of Chernobyl remains highly contested (Parfitt, 2006; Petryna, 2013).

2 Chernobyl Forum (2006).

3 Participants were asked if more safety measures are needed according to three categories: "**could be riskier**: it would be acceptable if it was __ times riskier"; "it is presently **acceptable**"; and "too risky to be acceptable: it would have to be __ times safer".

4 Exceptions: swimming, vaccinations, non-nuclear electric power generation.

References

Adams, J., & Hillman, M. 2001. The risk compensation theory and bicycle helmets. *Injury Prevention*, 7(2), 89–91.

Baum, A. 1993. Implications of psychological research on stress and technological accidents. *American Psychologist*, 48(6), 665.

Beardslee, W., & Mack, J. 1982. 'The Impact on Children and Adolescents of Nuclear Developments' in *Psychosocial Aspects of Nuclear Developments*. Report of the Task Force on Psychosocial Aspects of Nuclear Developments of the American Psychiatric Association, The American Psychiatric Association Task Force Report 20, Washington, DC: Author. pp 64–93.

Chernobyl Forum. 2006. *Chernobyl's Legacy: Health, Environmental and Socio-Economic Impacts*. Vienna: International Atomic Energy Agency.

Edelstein, M.R. 2004. *Contaminated Communities: Coping with Residential Toxic Exposure*. Boulder, CO: Westview Press.

Elliott, S.J., Taylor, S.M., Walter, S., Stieb, D., Frank, J., & Eyles, J. 1993. Modelling psychosocial effects of exposure to solid waste facilities. *Social Science & Medicine*, 37(6), 791–804.

Eyles, J., Taylor, S.M., Johnson, N., & Baxter, J. 1993. Worrying about waste: Living close to solid waste disposal facilities in southern Ontario. *Social Science & Medicine*, 37(6), 805–812.

Fischhoff, B., Slovic, P., & Lichtenstein, S. 1982. Lay foibles and expert fables in judgments about risk. *The American Statistician*, 36(3b), 240–255.

Fischhoff, B., Slovic, P., Lichtenstein, S., Read, S., & Combs, B. 1978. How safe is safe enough? A psychometric study of attitudes towards technological risks and benefits. *Policy Sciences*, 9(2), 127–152.

Flynn, J., Burns, W., Mertz, C.K., & Slovic, P. 1992. Trust as a determinant of opposition to a high-level radioactive waste repository: Analysis of a structural model. *Risk Analysis*, 12(3), 417–429.

Flynn, J., Slovic, P., & Mertz, C. 1994. Gender, race and the perception of environmental health risk. *Risk Analysis*, 14(6), 1101–1108.

Fried, J., & Eyles, J. 2011. Welcome waste–interpreting narratives of radioactive waste disposal in two small towns in Ontario, Canada. *Journal of Risk Research*, 14(9), 1017–1037.

Kahneman, D., & Tversky, A. 1984. Choices, values, and frames. *American psychologist*, 39(4), 341.

Kraus, N., Malmfors, T., & Slovic, P. 1992. Intuitive toxicology: Expert and lay judgements of chemical risks. *Risk Analysis*, 12(2), 215–232.

Kunreuther, H., & Easterling, D. 1990. Are risk-benefit tradeoffs possible in siting hazardous facilities? *The American Economic Review*, 80(2), 252–256.

Parfitt, T. 2006. Opinion remains divided over Chernobyl's true toll. *The Lancet*, 367(9519), 1305–1306.

Petryna, A. 2013. *Life Exposed: Biological Citizens after Chernobyl*. Princeton, NJ: Princeton University Press.

Rohrmann, B. 1994. Risk perception of different societal groups: Australian findings and crossnational comparisons. *Australian Journal of Psychology*, 46, 150–163.

Schneider, M., Froggatt, A., & Thomas, S. 2011. 'Nuclear Power in a Post-Fukushima World' in *The World Nuclear Status Report*. Washington, DC: World Watch Institute.

Slovic, P. 1987. Perception of risk. *Science*, 236(4799), 280–285.

Slovic, P. 1993. Perceived risk, trust and democracy, *Risk Analysis*, 13, 675–682.

Slovic, P., Fischhoff, B., & Lichtenstein, S. 1980. 'Facts and Fears: Understanding Perceived Risk' in Schwing, R., & Albers, W. (eds.). *Societal Risk Assessment: How Safe Is Safe Enough?* New York: Plenum.

Slovic, P., Layman, M., & Flynn, J.H. 1991a. Risk perception, trust, and nuclear waste: Lessons from Yucca Mountain. *Environment: Science and Policy for Sustainable Development*, 33(3), 6–30.

Slovic, P., Layman, M., Kraus, N., Flynn, J., Chalmers, J., & Gesell, G. 1991b. Perceived risk, stigma, and potential economic impacts of a high-level nuclear waste repository in Nevada. *Risk Analysis*, 11(4), 683–696.

Small, K.A., & Van Dender, K. 2007. Fuel efficiency and motor vehicle travel: The declining rebound effect. *The Energy Journal*, 28(1), 25–51.

Starr, C. 1969. Social benefit versus technological risk: What is our society willing to pay for safety? *Science*, 165(3899) (Sep. 19), 1232–1238.

Tversky, A., & Kahneman, D. 1974. Judgment under uncertainty: Heuristics and biases. *Science*, 185(4157), 1124–1131.

Whitfield, S.C., Rosa, E.A., Dan, A., & Dietz, T. 2009. The future of nuclear power: Value orientations and risk perception. *Risk Analysis*, 29(3), 425–437.

Wittneben, B.B. 2012. The impact of the Fukushima nuclear accident on European energy policy. *Environmental Science & Policy*, 15(1), 1–3.

WHO. 2002. *The World Health Report 2002: Reducing Risks, Promoting Healthy Life*. Geneva: Author.

5 Risk is everywhere

Introduction

As the world becomes more cosmopolitan – more focused in urban centres and increasingly interconnected – and environmental disasters seem to pervade the media, there is growing concern that environmental health threats have intensified substantially since the dawn of industrialization. Though technologies continue to emerge that are meant to minimize health threats there is a growing sense that the world is becoming less safe from harms that can be traced to human technology. This is despite the fact that life expectancy continues to increase in most jurisdictions (WHO, 2014). The spectre of increased global environmental health threats raises interesting questions for social scientists. Interest is not so much perhaps in determining *if* the world is on balance more or less dangerous due to technology; instead many social scientists want to know how debates about increasing environment and health threat themselves shape and are shaped by society.

The apparent ubiquity of risk brings into focus issues of scale; particularly the idea that environmental health threats are global. Though phenomena like global climate change, nuclear fallout and trans-boundary transport of air and water pollution are now firmly rooted in social consciousness through the media, acute and chronic disasters are still spatially uneven, felt more intensely in some places and less so in others. Thus, issues of environmental justice may actually be intensifying at various scales: globally between countries, but also regionally within countries and locally where specific technologies like chemical and nuclear facilities operate. For example, long-range transport of pollution is already being tracked at wide regional scales in places like Europe (UNECE, 2014), Asia (Liang et al., 2004), North America (Shen et al., 2005), across the Arctic (Vorkamp and Rigét, 2014) and even intercontinentally between Asia and North America (Jaffe et al., 1999). The substances being tracked range from carbon monoxide (Liang et al., 2004) and mercury (Durnford et al., 2010) to persistent organic pollutants (POPs; Beyer et al., 2000), all of which have potentially serious implications for both human and ecosystem health.

Trans-boundary contamination issues also apply to water. The by-products of industrial manufacturing, transportation, damming , human 'sewage' waste and the use of products such as pesticides and medicines all have the potential to find their way into local water courses, and in turn travel to jurisdictions well beyond where they enter the system. One particularly acute case is the Mekong River running from Tibet and China through Laos, Thailand, Cambodia and Viet Nam. Arsenic contamination of local wells and the river itself has recently been tied to natural processes but also anthropogenic practices that disturb sediments in the river bottom such as irrigation pumping, dredging and damming (Polizzotto et al., 2008). Increasing incidence of well contamination can be further tied to social processes such as growing population and the need for marginal populations in particular to find new clean sources of water amid water scarcity. Because arsenic is odourless and tasteless, well-meaning efforts of NGOs to create new wells for these communities have the potential, in the absence of proper testing, to exacerbate arsenic poisoning, a condition which can lead to skin lesions and cancer (Berg et al., 2007). In the Dniester River in Eastern Europe there are similar issues with contamination of sediments, but there pesticides and waste water are the problem more so than arsenic. In the Dneister sediment problems are combined with difficulties with the contamination of surface waters by various oils from industry along the river and its tributaries (Lebedynets et al., 2005).

Climate change is perhaps the best-known global-scale risk linked to technology to garner attention in recent years (IPCC, 2015). Climate change impacts likewise have a justice element with problems like sea level rise affecting mainly tropical coastal areas, while droughts, floods and severe weather events are somewhat more diffuse. The negative health effects in particular are not evenly distributed at any scale since poor and marginalized populations always tend to be the least resilient to disasters. Globally severe weather events disproportionately impact health in the developing world in the tropical regions. For example, the Sahel region stretching across northern Africa south of the Sahara Desert is both highly populated and highly vulnerable to droughts and famine that inevitably accompanies extended droughts. There is a long history of drought and famine in the region, one of the worst being in the 1970s to 1980s, but more recently in 2010 and again in 2012. In an area already prone to droughts, human-induced climate change has the potential to seriously negatively impact millions of people with only slight changes to this fragile ecosystem (Boyd et al., 2013). This situation has been worsened by religious and political violence, Darfur and Mali being two recent examples. Thus, global environmental hazards are intricately connected to social processes that put people at risk.

Similarly, coastal areas in the developing world are particularly vulnerable to severe wind events and coastal storm surges, yet because complex ecological systems are involved, there remains some debate about how much storm intensity and frequency have and will increase from anthropogenic climate change. For example, while Knutson et al. (2010) conclude through climate modelling that tropical cyclones will increase in frequency by 2 to 11% and in intensity by 6 to 34% in this century, in a review of the literature on tropical cyclones Pielke et al. (2005) suggest that the impact of increased cyclone activity will be relatively negligible. How things are measured can have an impact on what is to be done, whereby Webster et al. (2005) show that though the number of days experiencing tropical cyclones has decreased in recent years their intensity has increased. It is this type of uncertainty and debate over how to define the problem that provides fertile ground for theorizing about how such risks impact society broadly speaking and how society reacts through ongoing actions and institutional changes.

Some have suggested that our use of technology has not only changed society in terms of how we organize work, the city and the family, but that the environmental threats that accompany these technologies are dramatically changing or could seriously change the way we organize as a society – the very institutions on which societies have been traditionally based. How important is environmental health risk to the organization and inner workings of society as a whole? The risk society thesis developed by Beck and Giddens suggests that risk is becoming central to the way we think about our world and how we relate to each other as individuals and groups. They have suggested that because environmental threats are so globalized, risk is changing the way we organize as a society. For example, global conferences on the environment and dealing with climate change were unheard of only a few short decades ago at the same time the new institutions, like environmental impact assessment agencies, have emerged in many parts of the world. The motivation for such changes is the loss of *ontological security* – a sense of stability and continuity in everyday life (Giddens, 1991) – that was felt in the heyday of industrial development, when new technologies were seen mainly for their benefits in solving problems like increasing agricultural yield and providing cheap energy rather than for their potential harms. That is, globalized risk has awakened us to the idea that environmental harms have the potential to impact everyone, leading us to reflect on and think about ways we should be organizing society to better deal with technological threats. For example, traditional institutional arrangements like insuring against loss – the insurance industry (e.g., famine or toxic exposure insurance) – do not make sense in the era of global risk. Under the risk society old institutions evolve and new ones emerge.

The remainder of this chapter is organized largely around the risk society theory, from Beck in particular, because this theory has spurred one of the key moments in theorizing about environment and health in the social sciences. Along the way some critiques of the theory are highlighted and related theories of society in the age of risk are discussed, including Luhmann's systems theory as it relates to risk, Douglas and Wildavsky's cultural theory of risk and Kasperson's social amplification of risk framework. Before moving on to the details of Beck's thesis, how he defines risk is contrasted with more typical definitions to provide a window on what is to come in this chapter.

(Re)defining risk

It is worth reinforcing that risk society is not just about environmental health risks; it pertains to the idea that risks of all sorts are a feature of an increasingly globalized society. The risk society thesis provides commentary on a wide range of societal hazards ranging from crime, terrorism and the home to health, unemployment and environmental pollution, but the focus here will be health and the environment. For Beck, risk is socially and economically embedded, somewhat of a contrast to what is emphasized by most health risk managers (Chapter 9). For risk managers, definitions of risk revolve around calculating the threat of harm, which includes elements of both probability (How likely is it to happen?) and magnitude (How much damage could conceivably occur?). Yet for Beck risk is also about being socially and institutionally embedded:

> The concept refers to those practices and methods by which the future consequences of individual and institutional decisions are controlled in the present. In this respect, risks are a form of institutionalized reflexivity and they are fundamentally ambivalent. On the one hand, they give expression to the adventure principle; on the other, risks raise the question as to who will take responsibility for the consequences, and whether or not the measures and methods of precaution and of controlling manufactured uncertainty in the dimensions of space, time, money, knowledge/non-knowledge and so forth are appropriate.
>
> (Beck, 2000 p. xii; see also Mythen, 2004)

Four elements of this quote are worth emphasizing. First, the reference to "practices and methods" underscores that the manner in which we as a society define what is high and what is low risk is important. The latter part of the definition refers to the idea that we are in an era that causes us as a society to question how we determine what is safe (low risk), what is not and what we do about it. Second, Beck makes a distinction between individual and institutional decisions, hinting that these may not be entirely congruent in many cases. Individualization of risk is important for a number of reasons, in terms of how people make decisions to reduce risk but also how individuals and groups of individuals increasingly resist definitions of technological hazards risk that originate from institutions perceived to be old, out of date and incapable of adequately addressing risk. Third, the disagreements between institutionalized risk assessments and those of individuals are captured in the concept of reflexivity. This refers to the idea that when old ways of doing things do not work, institutions must pause, adapt and change. For Beck, those institutions that are already undergoing change are far reaching from political parties to the organization of the home, while the institutions involved in risk assessment and management are more directly involved in environment and health issues. Fourth, the combined elements of responsibility for consequences and uncertainty suggest that how we determine who has responsibility for safety from environmental toxins, for example, is becoming increasingly murky. In his most optimistic moments Beck sees this as a profound era for individuals and small groups to affect wide-reaching and positive institutional change, while in his more critical moments he suggests that because it is difficult to precisely apportion blame (e.g., to a negligent operator, manager or inexperienced firm), environmental damage may continue unabated. He labels this era 'reflexive modernity' and suggests we are at the precipice of profound social and institutional changes, ones we have not felt since industrialization.

The risk society theory

The main elements of the risk society theory were developed jointly by Ulrich Beck, a German sociologist, working closely at times with Anthony Giddens, a British sociologist, both of whom wanted to develop a theory to understand the transition from industrially based modernity to late modernity (Rosa et al., 2014). Volumes have been written about the risk society by Beck (1986, 1992, 1999, 2000, 2006, 2009), Giddens (1990, 1991, 1999) and others (Adam et al., 2000; Mythen, 2004; Richter et al., 2006; Zinn, 2008; Arnoldi, 2009; Rosa et al., 2014), but the focus here is on the core elements of the theory and how they relate directly to environment and health issues. The risk society is one of a range of social theories on risk covered in this book (e.g., cultural theory, social amplification of risk, psychometric risk perception), yet the risk society theory is perhaps one of the most elaborate and expansive in terms of scope, and it has received more popular attention outside academia and by professional risk managers than any of the other theories. As a result it is also among the most debated of the group.

One of the cornerstones of the risk society is an historical element that distinguishes eras of social, economic and cultural development. The central feature that differentiates one era from the next concerns the mode of distributing 'goods' (wealth) and 'bads' (by-products of production) in society, the latter being pivotal in the risk society era and relatively absent, or at least unnoticed, in previous eras. The eras are defined as pre-industrial society (traditional/feudal), industrial society (first/early modernity) and risk society (second/late modernity). Beck has been less concerned about precisely defining the years that delineate each era and more concerned about the conditions of each era that give us clues to the trajectory of society, that trajectory being defined largely in terms of how society deals with risk. What distinguishes traditional society from modernity is that the mode of production and distribution is centred on industrialization in the latter eras. Though first modernity extends back into the 1500s when society was traditional and organized largely around agricultural practices, late modernity is characterized by rapid growth in ideas and social change. Another key feature that distinguishes the traditional from the modern period is the growth of capitalism. Though industrial capitalism was born in what he defines as early modernity and continues into late modernity, one of the key features of the latter period is that capitalism has globalized and, according to the risk society, so too have technological hazards.

Technological hazards dominate risk society compared to natural hazards, which dominated societal concerns more so in previous periods. This distinction though is somewhat false in the sense that all hazards are socially defined. Decades of hazards research in geography and elsewhere takes issue with over-emphasizing the 'naturalness' of natural hazards. That is, the damage from natural hazards comes from the way societies organize to bolster resilience or exacerbate vulnerability – the latter often attributed to marginalized groups (Hewitt, 1983; Burton et al., 1993). However, Beck's purpose for focusing on technological hazards is that they are products of socially organized economic systems of industrial capitalism, which is layered on top of these vulnerability and resilience issues. Nuclear fallout, toxic chemicals, air pollution and genetically modified organisms are largely the products of human activities rather than hazards that people might experience in the absence of such systems. While hurricanes and tsunami maybe be more or less predicted and warned, Beck argues there is little that can be done to prevent them from happening outright, while for technological hazards, outright prevention is theoretically possible, particularly if the underpinning activity (e.g., nuclear power generation) is halted through socio-political decisions. In this sense, while there are elements of fate associated with natural hazards, fate need not be as prominent an issue for technological hazards. In this sense technological hazards are much more socially, politically and culturally defined.

Against this backdrop of distinct eras and an emphasis on human-produced technological hazards, Beck defines his thesis in terms of at least five core propositions that distinguish the risk society era of late modernity from the early modernity era when industrialization was just taking off:

- The negative impacts of production – the 'bads' (pollution) – have garnered far more attention.
- The scope and scale of negative impacts of industrial production are much greater.
- Risk has become more individualized in terms of who is expected to make decisions to protect safety.
- There is a loss of faith in experts to protect against the negative impacts of production.
- There is a rise in competing knowledge claims regarding environmental hazards.

The basis of these propositions is the numerous negative impacts of technology that have been inadequately addressed throughout modernity – particularly pollution. Further, the problem is said to be getting worse as we move deeper into late modernity, or at least the scope of the potential damage is now better understood. In the case of increased attention to environmental pollution Beck routinely cites a few specific examples like Chernobyl, genetically modified organisms, transboundary air and water pollution and more recently global climate change. Media coverage of environmental issues has increased since the early days of industrialization, and that coverage has paralleled growth in environmental organizations like the Environmental Defense Fund, Greenpeace, and Friends of the Earth. The work of Rachel Carson (*Silent Spring*) regarding the side effects of agricultural pesticides was an early example of issues of contamination gaining widespread media coverage (Neuman, 1990; Brown et al., 2001; Corbett, 2001). In this sense critiques of capitalism based on ideas of unemployment, flight of capital and labour versus management are argued to give way somewhat to environmental issues in the late modernity of the risk society. During the latter period environmental pollution coverage evolves to supplant labour issues partially because environmental issues have all of the same elements of human drama and interest these more traditional labour issues have had (Kitzinger, 1999). This was a time when environmental issues became more humanized. While early environmental movements concerned preservation of distant natural spaces mostly by the social and economic elite – conservationism – modern environmentalism grew to concern everyday people facing pollution. Critically, many environmental issues have evolved to be discussed mainly as human health issues (Burger, 1990) rather than as just preservation-of-ecosystem issues. Environmental pollution has indeed had serious health impacts on populations within early modernity – like the thousands of deaths caused by the Great London smog in 1952 to 1953 attributed to widespread pollution from coal burning (Bell et al., 2004). Yet it is the wide-scale coverage of a range of both acute and chronic environmental events (e.g., Chernobyl, global climate change) in more recent years that has prompted Beck to suggest that late modernity is qualitatively different.

One of the reasons industrial environmental hazards that have purported health impacts have garnered more attention in late modernity is that their scope is multiscalar in both space and time. Beck suggests that few escape technological hazard threats under globalization. He refers to the relationship between technology and negative impacts as having a *boomerang effect*; the risks involved are bound to impact not only the usual groups of poor and less privileged classes but the ruling classes as well. He has been criticized for this idea as global environmental injustices have intensified during late modernity (see Chapter 8). Though he does not dispute the idea of intensified injustices, what he is suggesting is that it is far more difficult to compartmentalize and localize technological hazard threats. One of Beck's examples of this is the 1986 Chernobyl disaster, which had radiation impacts that were international in scope (see Chapter 4). It is worth noting that Beck devised the risk society ideas prior to the disaster itself. Thus, in the risk society, pollution/contamination is increasingly recognized to have negative impacts that are global in scope, or at least such impacts do not stay neatly confined within the borders of nation-states. What this means for Beck in sociological terms is that the institutions that are predicated on nation-states are in need of reflexive change – institutional soul searching combined with a desire and capacity to adapt to changing global circumstances. In this sense, assessment of the Chernobyl nuclear disaster requires the ascendance of new international institutions capable of making similarly international decisions. At the same time environmental hazards like nuclear waste, nuclear meltdowns and global climate change have an insidious temporal element to them. The idea of sleeping catastrophes grows in late/second modernity according to Beck and is consistent with the ideas of Vyner (1988) and Perrow (2011), who claim that people are not only increasingly traumatized by the threat of 'invisible', odourless and tasteless threats; accidents – for example, at facilities handling contaminants – are actually 'normal'. As more of these accidents accumulate, public fear of future accidents grows (Beck, 1992; Perrow, 2011).

The individualization of risk goes hand in hand with the notion that social insurance schemes have already begun to break down. The social insurance mechanisms (e.g., the welfare state, personal insurance) in early modernity were built on the idea of sustaining long-term ontological security (Giddens, 1991). That is, insurance supports a feeling that everything is controllable and safe within certain parameters and that the 'bads' can be calculated. When things have gone wrong, financial and other damages have traditionally been compensated to restore the overall sense of security. Chapter 4 describes how human psychology related to gambles – that we are loss averse – plays into this form of social-institutional arrangement. In the second modernity ontological security breaks down as social insurance becomes untenable. According to Beck, there is no insurance against catastrophes that are potentially global in scope, like global climate change, trans-boundary radiation leaks and air and water pollution impacts from industries thousands of kilometres away originating in various countries. According to Beck, institutions that grew up with the welfare state must reflexively adapt to these new risk realities.

An important corollary to the decline of social insurance is that more burdens for safety fall on individuals, who increasingly find old social groups (e.g., labour unions) less meaningful. While early modernity was characterized by shared identity through common class consciousness, late modernity is characterized by an emphasis on the individual. Though this has opened doors to making positive healthy lifestyle choices involving diet, exercise and the use of 'safe' products, it has also meant that individuals/households are left to face complex risk decisions on their own. Personal security through behavioural adjustments has become central to apportioning responsibility for achieving positive health outcomes, while upstream causes are routinely ignored. Brown et al. (2001) provide an example of how this is played out, using the coverage of the causes of breast cancer in the media. They conclude that most of the coverage concerns individual behavioural lifestyle changes with scant mention of environmental causes and even fewer news pieces that pay attention to government and corporate responsibility for helping reduce breast cancer risks. Though individuals may have a certain amount of control over their own behaviours, even those too are socially reproduced; but the multi-causality of breast cancer leaves such health risks in murky limbo, uncertain and apparently beyond the public agenda. It is scenarios like this that are particularly worrisome according to Beck, as society may be doomed to continually reproduce their lack of attention to environmental health threats simply because our current systems of science and policy are not used to dealing with them. Beck nevertheless sees individuals and small groups as having the greatest potential to break the institutionalized incapacities of the risk society. Such ideas are consonant with Foucault's notion of governance – that in many ways individuals not only govern themselves but, through social interaction, govern each other (Foucault et al., 1991). The social acceptability of behaviour becomes a very powerful way to control actions – for example, smoking, using pesticides, eating genetically modified organisms, using alternative energy. It is through discourse (a collection of arguments that have widespread resonance among individuals and thus have the force to sustain a set of practices) that such social acceptability is propagated, repeated and reinforced to support the status quo until sufficiently convincing counter-discourses emerge. While governments play a role in messaging that supports the status quo, so too does private industry including the media. It is through counter-discourses that environmental health threats may be acknowledged and changed through grassroots movements, for example. A notion consistent with Beck's optimistic version of the future in the risk society. The discussion will return to the issues of individualization and institutional change after deliberating the roles of experts and competing knowledge claims.

What Beck and Giddens describe is a society increasingly attuned to technological hazards that in turn has implications for how people and institutions will rearrange to adapt. The industrial era was infused with optimism and a general belief that wide-scale movement of people to urban and suburban environments to work in manufacturing to mass produce goods represented positive economic and social progress. Though this period marked some rather profound challenges as cities grew and populations densified, the health professions seemed to keep pace with technologies (e.g., sanitation) and social/health arrangements (e.g., large-scale immunization programs) to increase life expectancy in the industrial countries moving into the 21st century. Yet Beck argues that there has become a growing *awareness* that the technologies that beget prosperity are accompanied by harms. Beck readily admits his bias towards the need to better control environmental threats (Beck, 2006). He suggests through the risk society thesis that he is not alone. Increasingly publics of various sorts are paying greater attention to health threats posed by industrial developments including chemicals manufacturing, petrochemicals refining, nuclear technologies and genetically modified organisms. That is, technologies are increasingly questioned at the local level and beyond because a local facility presents a pollution health threat or because an industry poses a more generalized global threat. The Love Canal incident described in Chapter 1 was an early example in North America. In that case, the greater awareness of the impacts of industry literally seeped into local consciousness through their basements from an old chemical waste repository. The Love Canal incident also brought into focus the fact that scientific experts could not always diagnose and resolve these threats to the satisfaction of those most directly affected by the pollution (Gibbs, 1997). This loss of faith in experts is one of the key social processes that Beck argues has set up a moment in late modernity when institutional arrangements have the potential to radically shift their focus away from systems that invest in unconditional and unquestioned faith in unbiased expert systems. He argues for the need to move towards more participatory institutions that rely on input from local and other publics, not just experts from within traditional institutions.

The limits of science and sources of uncertainty for connecting environmental threats to human health outcomes are central themes in both this book and Beck's risk society. For example, when a new facility is proposed for a community like the several attempts to site nuclear waste repositories in the U.S. (Kunreuther et al., 1990; Slovic et al., 1994) and Canada (Rabe, 1994; Eyles and Fried, 2012) or the contamination remaining from an old industry requires cleanup (Furimsky, 2002), there is often scientific expert disagreement about how to proceed. In a general sense then, the science that resulted in seemingly unfettered progress and helped protect us unequivocally from harm in early modernity is seen to falter as a resolute basis for ensuring safety as late modernity takes hold. The various limitations of environmental health sciences like epidemiology and toxicology are reviewed in Chapters 1 and 3, and Beck's thesis picks up on how the various forms of scientific uncertainty play out in the everyday lives of people trying to get on in the world. The answer to questions like, "Does this facility pose environmental health risks?" are not simple ones, and this is the reason for the upheaval Beck describes. For example, in a case of the contamination of cattle feed by a flame retardant (PBB) in Michigan, Reich (1983) argues that the scientific uncertainty about the health impacts to humans from contaminated milk products led to the merging of politics and science to address the various problems posed by the technological disaster. Similarly, recent debates over the health impacts of energy technologies ranging from natural gas fracking (Mitka, 2012) to hydro dams (Lerer and Scudder, 1999) and wind turbines (Shepherd et al., 2011) are waged in a context where there has been scant scientific (epidemiologic) study specifically on health effects near such facilities. Though there may not be a total loss of faith in science, such situations often leave the average citizen wondering which scientists to believe. This also casts a spotlight on the politics behind creation of scientific evidence within changing institutional arrangements.

Late modernity is thus characterized by what Funtowicz and Ravetz (1993) call "post-normal science", whereby there is so much uncertainty that scientific experts and scientific evidence are no longer expected to dominate decision making. They too do not suggest science is being ignored entirely, merely that in certain contexts its status becomes more ambiguous, allowing other decision makers room to be involved. In the process, the many value judgements that seep into scientific assumptions are exposed and examined. Funtowicz and Ravetz characterize decision making within post-normal science as a function of two key dimensions: the *stakes* involved and *system uncertainties*. When both are low, science remains largely uncontested, and scientific experts can be left to apply their scientific knowledge unfettered; however, when both decision stakes and uncertainties are high controversy tends to be elevated due to the fact that impacts and outcomes remain contested. It is in the latter context that Beck argues institutional change will continue to be shaped – for example, to better include various publics.

Thus, in Love Canal even when a health study was completed (Heath et al., 1984), because the study focused on chromosomal changes that may or may not lead to cancers, there remained high levels of uncertainty about what the real health impacts to locals would be and, in turn, what actions should be taken. Ultimately, a precautionary approach was taken and residents were evacuated, but not before much local and wider activism drew attention to the issue (Gibbs, 1997). Such situations, Funtowicz and Ravetz argue, call for mobilizing an extended community of decision makers, an issue taken up more fully in Chapter 9. Post-normal science, they argue, suggests a prominent but humbled role for scientists, and decision makers need to more readily admit how and when they draw on social values (Latour and Woolgar, 2013). Whereas applied science has been viewed in the past to be value free, in the era of post-normal science, values play a much more prominent role in policy and action and precautionary approaches. Beck and Giddens argue that looking at science in this way may again assure safety despite the fact that there is always need for more data and more research. The need for more science need not mean accepting the status quo of uncertain (potentially high) risk.

Yet in the context of the risk society theory it may be argued that even problems that were once considered issues of 'applied science' have now become contested problems that are now within the realm of post-normal science. This is due to increasing awareness of the limits of science generally, regardless of whether scientists feel they have come to a consensus in a particular domain. The fluoridation and chlorination of water systems are examples of relatively newly contested remnants of long-standing applications of applied science. For example, the Fluoride Action Network (2014) keeps track of the growing number of communities that have voted against a decades-old practice of fluoridating public water systems in the name of reduced costs to health systems from dental caries. These votes against fluoridation suggest that counter-discourses can be effective for changing public policy in the name of protecting public health – no matter how remote the risks may be.

Beck also points out that there is increasing awareness of what we do not know. That is, we have become more attuned to our own ignorance in terms of what might go wrong (see also Perrow, 2011). For example, the British Petroleum disaster in the Gulf of Mexico that killed 11 workers and spilled millions of gallons of oil into the ocean was not a known outcome of any officially anticipated system failure. If industry officials had known, presumably there would have been systems in place to mitigate the problem much quicker. The infamous quote from former U.S. Secretary of Defense Donald Rumsfeld regarding the weapons of mass destruction in Iraq summarizes in an almost comedic way what we face today in terms of risk decision making:

> As we know, there are known knowns; there are things we know we know. We also know there are known unknowns; that is to say we know there are some things we do not know. But there are also unknown unknowns; the ones we don't know we don't know.
>
> (Rumsfeld, 2002 as cited in Rosa et al., 2014)

Science and policy have traditionally been based on probability and margins of error that are statistically knowable, 'known unknowns', while in the risk society it is the disasters that have not even been anticipated – 'unknown unknowns' like the BP Gulf of Mexico disaster that increasingly worry the public. Further such catastrophes identify a vast policy gap in terms of ways to decide the appropriate courses of action.

Beck cautions that there is more than an erosion of faith in science for unilaterally solving problems at play in decisions within the risk society. Though science is certainly part of how people form views on hazard threats, risk itself is socially constructed in the domains of daily life, information consumption, work, family and politics. For example, the social amplification and attenuation of risk framework forwarded by Kasperson et al. (1988) is known for highlighting the role of the media, for example, in social processes that serve to heighten (or attenuate) concern about various hazards (see Chapter 9). It may be argued that the relatively recent information age of fairly unfettered access to material on hazards and facilities of various sorts has served to intensify processes of amplification. It is in this context, Beck argues, that society is fundamentally changing in relation to and as a result of technological hazard risk. In fact, he argues that health threats from technological hazards are already changing the institutions that undergird industrial capitalist societies:

> After 100 years of radical modernization, the different dimensions of risk have not only come to pass, they are changing and threatening the very foundations of modernization as we know it – or, to be more precise, the basic institutions of first modernity: the core ideas of the nation state, the nuclear family, class conflict, international relations, the welfare state, nation-state democracy and scientific knowledge.
>
> (Beck and Lau, 2005, p. 550)

In the risk society the preoccupation with risk crosses old class boundaries. Workers and owners of capital are said by Beck to be equally exposed to globalized risks like climate change, genetically modified organisms, nano-technology, nuclear threats and terrorism. This has implications for social organization whereby group identities must necessarily shift. While working-class labour continues to push for a fair wage, this is further complicated by the fact that the same sectors that provide them with jobs also pollute their backyards and potentially threaten their health. That is, there is an erosion of class identity towards identities concerned about threats to individual health (e.g., GMOs) or localized conflicts over the definition of risk (e.g., locating a technology locally like GMO plantings). Not being able to rely on familiar social groups to handle such problems, Giddens argues, chips away further at ontological security within the risk society. Fighting managers for better wages is a different type of battle than fighting industry and government for cleaner, healthier environments. People are faced with the destabilizing realization that the groups to which industrial societies have traditionally turned for action against all manner of threats (e.g., labour unions) are not as well equipped to define actionable change as they were in the industrial era. Further, there are often competing sets of knowledge claims and recommendations for action concerning environmental risk as people struggle to graft old political views – for example, left versus right – into new amorphous, multi-causal environmental health problems.

The risk society theory and other theories of risk

Due in part to the popularity of the risk society theory a number of critiques have emerged, largely to enhance theoretical development. The first concerns engagement with other theories of risk, keeping in mind that the risk society is a grand theory of society rather than a theory meant to understand technological hazard risk alone. Dean (1999) is concerned that Beck and Giddens have not seized on opportunities to link the risk society theory more thoroughly to other broad sociocultural theories of risk (see also Lupton, 1999). In fact, Dean laments that overall there is very little engagement between some of the key sociocultural risk theorists – for example, risk society theorists Beck and Giddens have rarely engaged with Douglas and Wildavsky's (1982) cultural theory of risk. While the risk society is focused on global scale of hazards, Douglas and Wildavsky (1982) instead focus on the idea that social groups tend to emphasize a few environmental threats over all others. They endeavour to understand why the hazards laypeople emphasize are not necessarily the ones experts would suggest are a priority according to available data. For Douglas and Wildavsky (1982), what is essential to understand is both why we tend to narrow our attention on certain hazards and not others and the purpose that this narrowing of attention serves. While Beck and Giddens concentrate more on the de-stabilizing social realization of the limits of science and scientific experts, Douglas and Wildavsky argue that the choice of hazard for attention is central, since this is connected to the way of life practiced by a group. For example, when a way of life is interpreted to be threatened by technology – such as the proposal of a new facility in a community – the affected community is expected to react very critically. Though Boholm (1996) is concerned that way of life is rather poorly defined in this literature; Douglas and Wildavsky submit that it concerns attitudes, beliefs, and ways of interacting within a society. Agricultural ways of life in rural communities may be threatened not only by the chemicals that may leach out of facilities and into crops and livestock, but the act of putting the facility there itself may threaten the social way of life resulting in community upheaval (Baxter, 2006).

Rosa et al. (2014) further outline confluences in and divergences between key risk theories. They borrow from Funtowicz and Ravetz's post-normal science ideas to map the ontological foundations of various social frameworks for understanding risk (Figure 5.1). Like the original diagram, the ontological foundation is said to depend on the type of risk environment that is under consideration. According to Rosa et al. the environment for knowledge claims is at the heart of how we might theorize risks. When overall system uncertainty and decision stakes are low the main stakeholders are experts using traditional scientific risk analysis (see Chapter 2). Environmental assessment for routine construction projects such as highways, bridges and urban infill would fall into this category. In this context claims made by experts according to their internal rules of assessment are left relatively uncontested. In contrast, when the environment for knowledge claims involves high system uncertainty and high decision stakes (e.g., oil pipelines, terrorism and drought exacerbated by climate change), then there is greater latitude for treating knowledge claims as social constructions. That is, stakeholders may find it difficult to find common ground when assessing the acceptability of risks and may focus on distinctly different aspects of or types of threats. For example, in the environmental assessment process for a nuclear waste repository proponents may point out the rather impressively low death rate in the nuclear power industry over the last several decades (see Chapter 9), while local residents may point out the lack of specific data on leakage safety for the storage system being imposed.

Rosa et al. use the concepts of *ostensibility* (pointing to examples of safe operation) and *repeatability* (pointing to repeated instances of safe operation over time and space) to distinguish the movement from the origin outwards in Figure 5.1. Routine construction projects are high on both dimensions and are thus located toward the origin in the diagram where decision stakes are low, while nuclear waste repositories are low for both ostensibility and repeatability and are therefore located away from the origin where decision stakes are high. For example, there are not many deep geologic nuclear waste storage facilities worldwide, and the ones that are operational have only been working for a short period of time (World Nuclear Association, 2014). Synthetic realism is a middle ground for risks and theorizing about them and is an area occupied by several technologies. These are the ones with which we have experience, but they are nevertheless open to catastrophic failure in rare instances and are often the sites of social conflict over the imposition of risk. Nuclear power plants, pipelines, chemicals manufacturing

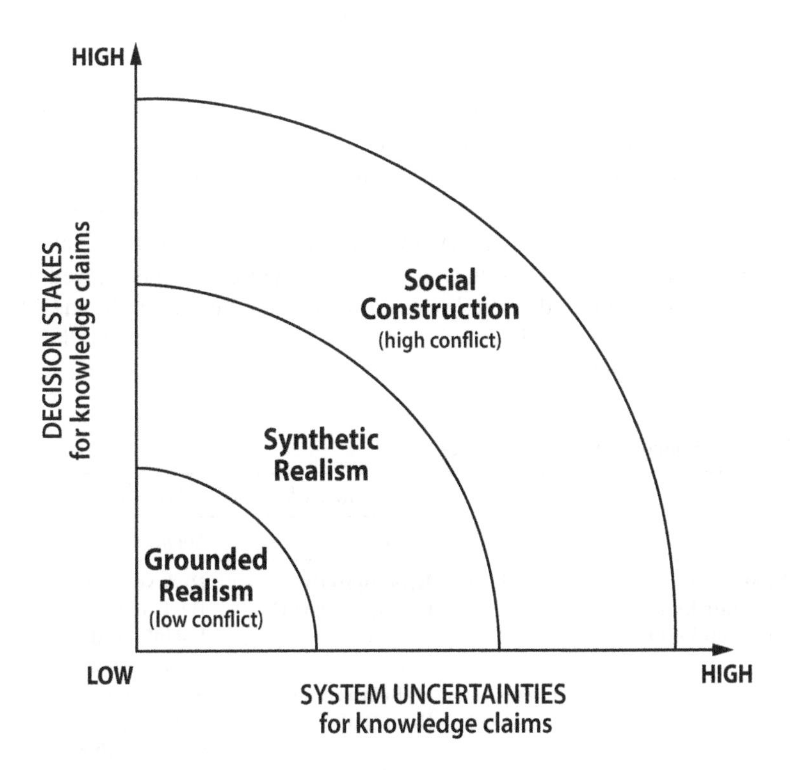

Figure 5.1 Ontological foundations of social theories of risk

Source: Adapted from Rosa et al. (2014).

facilities and offshore oil production fall into this category. Though experts can point to numerous examples of safe operation (high ostensibility) over long periods of time (repeatability), various stakeholders including those living close to proposed new facilities tend to point out one or both of two main categories of objections. First, there may be disagreement over the meaning of the existing data. Though deaths and illness may be exceedingly rare, how we die matters, and dying from radiation or chemical exposure is particularly distasteful. Further, exposing some and not others without compensatory benefits adds elements of injustice. Second, stakeholders are also prone to emphasize the seriousness of rare events. Chernobyl, Fukushima and the 9/11 attacks are all instances of events that awoke people to the 'reality' of rare events. That rare events have actually now happened is in fact one of the main contexts for increasing efforts at theorizing about risk in the first place.

Table 5.1 summarizes some of the main aspects of the risk society theory by juxtaposing it with other social theories of risk discussed throughout this book. The rows are defined according to epistemology – what is considered legitimate knowledge – while the columns are defined according to ontology – the way the world is assumed to exist for social scientists to study it. These dimensions are elaborated further in what follows, but for orientation formal risk analysis, described in Chapter 2 and located in the upper left cell, is based on the assumptions that knowledge should be objective by minimizing the role of social values and should focus on measuring a single, more or less invariant reality. This is why risk analysis is often distinguished from the more value-laden (subjective) aspects of risk communication and risk management. Latour and Woolgar (2013) caution, though, that the simplifying assumptions and choice of what precisely to study in risk analysis also involve (often less visible) value judgements. The social amplification of risk framework, sharing a cell with the risk society in the upper-right cell is elaborated in Chapter 9. The hybrid model advanced by Rosa et al. (2014) and located in the lower left cell, is predicated on the idea that we can never know when we have *adequately* measured and understood the dangers of a technology in the 'real' world, and we must therefore use social science to

Table 5.1 Foundations of key theories of risk

		Assumptions about the reality of risk (Ontology)	
		Realist	*Social construction*
What counts as legitimate knowledge (Epistemology)	Objectivist	**Risk analysis** – traditional expert analysis	**Risk society theory** – Beck and Giddens **Social amplification** – Kasperson et al.
	Subjectivist	**Hybrid risk model** – Rosa et al.	**Systems theory** – Luhmann **Cultural theory of risk** – Douglas and Wildavsky

Source: Adapted from Rosa et al. (2014).

understand the subjectivities surrounding risk. Yet they suggest that all subjective positions (claims to knowledge) are not and should not be created equally. They are concerned that opening up too much to a subjectivist/constructivist paradigm may be interpreted to advocate that good policy can be based on any set of knowledge claims that gains hegemony in policy circles – regardless of the physical scientific basis of those claims. In this sense, there are presumed to be good and bad knowledge claims based on strong versus limited science; and for Beck and Rosa et al., despite slightly different locations in the table, the problem then is sorting out which knowledge claims ascend to widespread public favour and why.

The hybrid model contrasts Luhmann's (1989, 1993) systems theory, which, like the risk society theory, is a grand theory of society for which risk plays a central role. His main thesis is that society is made up of multiple systems that are self-sustained by defining their own 'reality' and image of outsiders in other systems – that is, such systems are *autopoietic*. Society is thus comprised of different social groups making their own knowledge claims that are bound, at certain points, to conflict. In this schema, those systems that subsume a technology favourably into their own system (e.g., trans-boundary water transfers, coal or nuclear power) will come into conflict with those that do not (e.g., water sovereignty, wind and solar power). In contrast to Rosa et al.'s hybrid model, there are no right and wrong systems per se, only those which generate discourses sufficient to put them in positions of power to enact their policy preferences and those which pursue counter discourses to the ones of those in power (see also Foucault et al., 1991). Risk science, then, is used as one of many tools to support those discourses. While the hybrid risk position and traditional risk analysis suggest that risk science should categorically be central to decision making, within systems theory, Luhmann proposes that such normative positions simply signal where one academic or another is already placed within the set of systems. He suggests that we need to pay much more attention to the socio-politics of decision making under current institutional structures. This emphasis on socio-politics and institutional decision making aligns well with Beck's risk society ideas, but Beck (2006) is more future-oriented and even prescriptive in terms of delineating why and how institutional change happens.

The cultural theory of risk, advanced by Douglas and Wildavsky (1983), is complementary to Luhmann's systems theory but with less emphasis on discourse and more emphasis on the moral/philosophical underpinnings of various groups in society. It is worth noting that the book *Risk and Culture* came to the English-speaking world prior to the works of Beck, Luhmann and Giddens. Nevertheless, these approaches share the perspective that the social constructions of risk and group subjectivities are central in thinking about risk. The cultural theory of risk builds on the ideas developed by Douglas (1966) while studying the Lele people of Western Africa to understand how and why social groups maintain social order through shared worldviews. These ideas have been distilled into two dimensions – grid and group – which circumscribe four main worldviews (see also Tansey and Rayner, 2009). *Grid* refers to the degree to which a society or way of life emphasizes difference and stratifies into roles or, instead, emphasizes sameness and the idea that roles are largely interchangeable. *Group* concerns the degree of social control with a continuum ranging from self-sufficiency to complex rules governing social actions. The combinations of grid and group depicted in Figure 5.2 circumscribe both ways of life and the worldviews that support particular ways of life. Like the other theories in the right-hand side of Figure 5.2, in the cultural theory of risk these ways of life may conflict with each other within broader society.

The egalitarian worldview is characterized by risk aversion, a desire to guide risk decisions using moral and typically precautionary principles. Within Western societies in particular those who align with this worldview are concerned about catastrophe from system failures, which parallels the risk society ideas about loss of faith in modernist expert systems and institutions. Hierarchist and individualist worldviews tend to support the idea that we have not taken on too much risk; indeed, an individually oriented worldview supports risk taking and emphasizes the use of unfettered free markets to simultaneously ensure profits, safety and conflict avoidance. The hierarchist worldview on the other hand is buoyed by good bureaucracy and the assumption the best way to manage safety and conflict is through authoritative control. The struggles over risk in developed capitalist countries tend then to be dominated by power struggles between these latter three

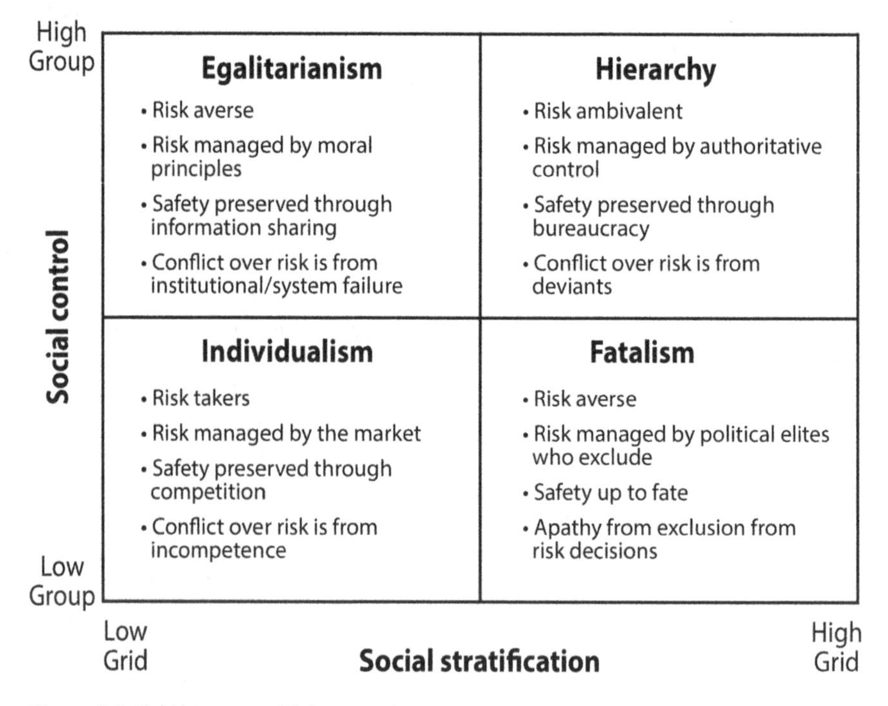

Figure 5.2 Grid/group worldview typology

Source: Adapted from Douglas and Wildavsky (1983).

ways of life. The fatalist worldview on the other hand is associated with the marginalized and disenfranchised in Western democracies. They tend to be minority groups with little access to power structures, and though they are risk averse, they tend to feel that the ability to control safety is so far out of reach that they leave such matters to fate. Douglas and Wildavsky used this schema to suggest that up to the 1980s, institutional change concerning the environmental protection was driven by egalitarian-oriented environmental organizations squaring off against both individualist-oriented industry and hierarchy-oriented governments and government institutions (e.g., environmental assessment). Douglas and Wildavsky further argued that a main effect of these conflicts was to further solidify the ways of life of each of these groups but at the same time concede that the interplay between groups has an impact on institutional change (see also Tansey and Rayner, 2009). Thus, cultural theory shares a number of elements with Luhmann's systems theory in the sense that groups with different views on hazard threat create social space in which institutional change may be encouraged.

Returning to Table 5.1, the main ontological question is whether there are real risks or, more to the point, real threats/dangers that exist independently of our social understandings of them. This is a question poised on the precipice of a very slippery slope, in the sense that suggesting that phenomena like nuclear radiation and its link to cancer are not 'real' seems absurd, indeed offensive, to many. This is why Table 5.1 is best understood alongside Figure 5.1 since the latter further elaborates the role of ontology in terms of decision stakes and system uncertainty. The ontological question around risk might better be framed as, "*What are the opportunities for multiple competing interpretations of what is important for decision making?*" In this sense, what is 'real' is what is important for one group or another and hence is actionable. When stakeholders disagree over imposed risks it is not necessarily the 'reality' of the numbers that is in dispute – for example, the number of people who died because of the Chernobyl disaster – it is the circumstances and meaning of those deaths and other aspects of institutional arrangements (e.g., broad issues of justice) that distinguish one set of stakeholder knowledge claims from another. When professional risk analysts remind of the 'reality' that people are more apt to die from their morning routine getting to work or school than they are to die of radiation or chemical exposure, it is not the only reality worthy of attention when the meaning of chronic, debilitating disease imposed by the actions of distant others, however unlikely the scenario, is explored more fully. In this sense those theories that emphasize social constructions are doing so not to deny what one group or another of experts reveal in their calculations per se, it is to suggest that the meaning of what they find – the danger – is what motivates conflict over knowledge and policy. So when these theorists assert that social constructionism is based on the idea of multiple parallel realities, those realities are based largely in meaning and making sense of the system of risk – be they nuclear technologies, GMOs, climate change, chemicals or some other human-made hazard.

In contrast to the right column of Table 5.1, traditional risk analysis and the hybrid model proposed by Rosa et al. focus instead on the idea that understanding the reality of the dangers, the numbers and the probabilities is paramount and that everything else should or at least could flow from that research. That said, the hybrid model, as its name suggests, is more attuned to the idea that risk policy is embedded in socio-political structures that can dramatically impact decision paths regardless of what the science might suggest. Despite where it is depicted in the figure, this approach may serve as a potential bridge between the realist approach of basic risk analysis and the social constructionist approaches preferred by many social scientists.

The epistemology dimension in Table 5.1 distinguishes differences in terms of what are considered legitimate knowledge claims. Objectivity is based on the idea of minimal human interference. For example, an epidemiologic study of the health impacts of environmental noise would favour blood and urine samples to measure stress and wearable monitors to measure sleep quality. For example, Babisch et al. (2001) conclude that the amount of night-time traffic noise is correlated with elevated catecholamine levels in urine, particularly for those who could not sufficiently reduce the noise disturbance by closing a window. Less objective measures in such a study would be survey responses to questions that form a stress scale, while quotations from interviews with exposed residents talking about the stresses in their lives would be considered relatively more subjective. In a different study of the impact of environmental noise on stress, Wallenius (2004) uses a questionnaire with several scales including the Personal Project Inventory, the EPI scale of neuroticism, the Somatic Symptom Checklist and questions about general health status, noise annoyance and daily activities instead of objective somatic measures. The advantage of this approach is that the impacts on daily functioning are placed in the forefront.

However, survey and interview-based measures raise questions about legitimacy and bias. Put more simply, "Are subjectivist positions on risk legitimate enough to be worthy of our theoretical and policy attention?" For many the answer to this question is yes, partially because the concept of objectivity as value-free measurement has been heavily criticized. Some measurement instruments may not be sensitive enough and may have a tendency to lead to false negative results (showing no effect when there would be with other instruments). Good scientists report such instrumentation biases if they are known, yet the results of such studies are often reified – take on a life of their own – despite such errors. Latour (1999) suggests that potentially value-laden distortions in conceptualization and measurement happen throughout all scientific endeavours. For risk theorists like Luhmann (1993) this does not signal the need to abandon science but instead to open the door further to subjectivities of risk, since it is these subjectivities that will guide how traditional scientific risk analysis is interpreted by lay publics including decision makers. As Thomas and Thomas have suggested (1928), what people *believe* determines how they will act regardless of whether what they believe is 'correct' according to some scientifically agreed upon standard. In this sense those attuned to the subjectivities of risk focus attention on why people hold fast to particular knowledge claims and not others, even in the face of contrary evidence. Subjectivities may also involve defining what aspects of a hazard are the most risky, that is, most important and most worthy of mitigation.

While those on the top row of Table 5.1 have faith that the core principles of objectivity are salvageable and sustainable, those on the bottom tend to feel the foundations of objectivity have crumbled enough that we need to pay greater attention to other ways of knowing. Whether Beck's risk society theory belongs in the objectivist or subjectivist group is likely debatable. Beck has been pressed several times for his position on this issue, and like Rosa et al. (2014) he agrees that there is a key role to be played by (objective) risk analysis (Beck, 2006). This is somewhat of an ironic twist in the sense that it is the objectivity of risk analysis that Beck argues has caused reflexivity in late/advanced modernity. The capacity of objectively based risk analysis to both raise expectations of safety and fail to fulfil them is what has caused society to be more critically aware of science and the institutions supported by science. For Beck, then, it is that objectivity that is driving reflexive institutional change. Likewise Kasperson et al.'s (1988) social amplification of risk framework is predicated on the idea that there is a start point from which risk is either amplified or attenuated in the public consciousness. That start point is presumably expert consensus on objectively derived risk calculations which may nevertheless be opened up to debate by fearful publics.

Institutional change in the risk society

Many of the theories discussed deal in one way or another with institutional change from risk. Giddens, and particularly Beck in the risk society, have suggested that because the foundations of our understandings about technology have been so irrevocably shaken, institutional changes are both inevitable and desirable. Since the risk society focuses on globalized hazards, we highlight a few notable examples of international policies: the Montreal Protocol on Substances that Deplete the Ozone Layer (1987; UNEP, 2014), the Great Lakes Water Quality Agreement (GLWQA; 1972; Environment Canada, 2014a, 2014b), the Rio 'Earth Summit' – United Nations Conference on Environment and Development (UNCED) United Nations, 2014) and the Kyoto Protocol (1997; UNFCCC, 2014). There has been mixed success from these international agreements, with the Montreal Protocol and Great Lakes Water Quality Agreement generally applauded, while the Rio Earth Summit and the Kyoto Protocol tend to be more heavily criticized for failing to live up to relatively ambitious goals (Wynne, 1992; Jasanoff, 2002; Renn, 2008).

The Montreal Protocol's success has much to do with the level of agreement on the science of chlorofluorocarbons (CFCs) and atmospheric ozone. That is, there is low scientific uncertainty about the need to reduce chlorine and bromine-containing CFCs to regenerate atmospheric ozone and reduce the ecosystem impacts of health-harming ultraviolet radiation. There was a general consensus among scientific and lay communities that something needed to be done. Importantly the recommended actions did not threaten any group enough to mobilize convincing counterclaims (Velders et al., 2007). One of the reasons for this may be that for many CFCs there were non-CFC substitutes – for example, for fire retardants and aerosol can propellants – available from the same chemical producers.

Thus, there was little impact on individual lifestyle in terms of changing behaviours and little change in the ability of industry to remain profitable in both the short and long term (Murdoch and Sandler, 1997).

The Great Lakes Agreement involves a wider variety of issues and arguably more complex ecosystem science than in the case of CFCs and the ozone layer. Nevertheless, in this case there were only two nation-states involved – Canada and the U.S. Thus, issues of historical colonialism under capitalism and a whole host of other historical socio-political and economic issues that dominated Rio and Kyoto were not as prominent in the GLWQA discussions. The goal has been to reduce chemical pollution in the Great Lakes to better preserve ecosystem integrity. Among the reasons claimed for the success of this arrangement is the history of close ties between the two countries and a sense of common purpose and mutual benefit. As well, the long history of the GLWQA has allowed for extensive consultation not only between scientists but between scientists and other interested stakeholders such as local residents including First Nations (Environment Canada, 2014b). This process has also produced new forms of institutional arrangements – the most central being the International Joint Commission (IJC, 2014), which has transcended national government structures with the purpose of managing trans-boundary pollution. In Beck's thesis, such adaptive institutional arrangements are expected for dealing with new global problems as societies reflexively reorganize; but achieving similar success on such broad environmental goals on a more global scale has proven to be significantly more challenging.

The Kyoto Protocol – which is an international agreement to limit greenhouse gas emissions – has been far more heavily criticized than either the Montreal Protocol or the Great Lakes Agreement. The Kyoto protocol took shape at the Rio summit five years earlier, the latter being largely a forum for non-binding intentions regarding environmental sustainability including social justice. Like the Montreal Protocol, Kyoto is global in scope, the difference being that substituting for activities that are contributors to greenhouse gas emissions (e.g., energy production, transportation and heavy industry) is far more onerous than it is to substitute for the CFCs. In fact, since the Montreal Protocol was so heralded as a model approach to international pollution control agreements, the Kyoto agreement was based on that earlier process. The main parallel was the use of emissions targets – for example, methane and carbon dioxide – according to a specific timetable. Some have been critical of this idea, suggesting that nobody had taken the time to determine if there were sufficient parallels between the two emissions problems to warrant similar approaches (Victor, 2004). Though many signatories to the Kyoto accord have met their interim targets, overall greenhouse gas emissions rose substantially and consistently from 22.7 billion tonnes of carbon dioxide equivalent greenhouse gases in 1990 to 33.9 billion in 2011 (Schiermeier, 2012). One of the problems has been the thorny issue of justice, coming up with a system to allow developing nations to economically lift large portions of their population sustainably out of poverty. Some of those same critics have put forward counter-claims for a different policy path and recommend a system based less on binding targets and more comprehensively on *emissions trading* to get developing nations more thoroughly involved (Victor, 2004).

Though there is much to be applauded from Rio to Kyoto in terms of international cooperation, Beck's vision for a society reflexively generating institutions to handle complex problems like climate change is not yet being fully realized. The limited success with global greenhouse gas emissions, though, fits nicely into the heuristic space for theorizing about such problems. Global climate change likely fits in the post-normal science region of Figure 5.1 and the social constructionism regions of Table 5.1. What this means is not that traditional sciences have no value; on the contrary, it means that we must consider how that science is socially constructed within various groups within society. This does not suggest a supplanting of science by social science; rather, this kind of thinking is meant to highlight the importance of new layers of understanding to deal with these complex problems. By way of contrast, bi-lateral agreements between similarly developed economies as is the case of the Great Lakes agreement or multilateral international agreements based on *relatively* narrowly scoped problem easily substituted substances (CFCs) as in the Montreal Protocol both seem to occupy a space somewhere high on the decision stakes axes in Figure 5.1 but low in scientific uncertainties and conflict over knowledge claims. Though the disputes against the core scientific evidence of human contributions to climate change have died down in recent years, the debate over appropriate policy mechanisms and institutions to implement those mechanisms remains high.

Conclusions

The idea that 'risk is everywhere' has had a rather profound impact on how social scientists theorize about technological hazards. Though the risk society is likely the best known of these theories, the cultural theory of risk, social amplification of risk and systems theory are among the range of approaches to understanding how society is organized by and organizes to manage threats from human-created technological hazards like chemical pollution, acute contamination, radiation accidents and human-induced aspects of climate change. Nevertheless, these theoretical approaches share many assumptions in common including that (i) scientific uncertainty has eroded public confidence that science will readily suggest policy paths, (ii) amid this uncertainty groups have formed to advance competing knowledge claims in support of various policies to handle these often complex technological problems, (iii) the claims and policy recommendations, though based in science, have socially constructed elements and (iv) only by embracing the idea that knowledge claims and recommended policy paths are at least partially socially constructed can we hope to reflexively change policies and institutions to handle these hazards. Beck has suggested that as we move deeper into late modernity we will reflexively build a society increasingly around technological hazard risk. In the next chapter we will explore such ideas further as we address the question "Is the risk reversible"? Using more examples, we further discuss how risk policy challenges are being met to minimize risk amid uncertainty and to learn from conflict and public concern yet simultaneously address broader issues like social justice.

References

Adam, B., Beck, U., & Van Loon, J. 2000. *The Risk Society and Beyond: Critical Issues for Social Theory*. London: Sage.

Arnoldi, J. 2009. *Risk, an Introduction*. Cambridge, UK: Polity.

Babisch, W., Fromme, H., Beyer, A., & Ising, H. 2001. Increased catecholamine levels in urine in subjects exposed to road traffic noise: The role of stress hormones in noise research. *Environment International*, 26(7), 475–481.

Baxter, J. 2006. Place impacts of technological hazards: A case study of community conflict as outcome. *Journal of Environmental Planning and Management*, 49(3), 337–360.

Beck, U. 1986. *Risikogesellschaft: Auf dem Weg in Eine andere Moderne*. Frankfurt am: Suhrkamp.

Beck, U. 1992. *Risk Society: Towards a New Modernity*. London: Sage.

Beck, U. 1999. *World Risk Society*. Cambridge: Polity Press.

Beck, U. 2000. 'Forward' in Allan, S., Adam, B., & Carter, C. (eds.). *Environmental Risks and the Media*. London: Routledge. pii.

Beck, U. 2006. Living in the world risk society: A Hobhouse Memorial Public Lecture given on Wednesday 15 February 2006 at the London School of Economics. *Economy and Society*, 35(3), 329–345.

Beck, U. 2009. *World at Risk*. Cambridge: Polity Press.

Beck, U., & Lau, C. 2005. Second modernity as a research agenda: Theoretical and empirical explorations in the 'meta-change' of modern society. *The British Journal of Sociology*, 56(4), 525–557.

Bell, M.L., Davis, D.L., & Fletcher, T. 2004. A retrospective assessment of mortality from the London smog episode of 1952: The role of influenza and pollution. *Environmental Health Perspectives*, 112(1), 6.

Berg, M., Stengel, C., Trang, P.T.K., Hung Viet, P., Sampson, M.L., Leng, M., & Fredericks, D. 2007. Magnitude of arsenic pollution in the Mekong and Red River Deltas – Cambodia and Vietnam. *Science of the Total Environment*, 372(2), 413–425.

Beyer, A., Mackay, D., Matthies, M., Wania, F., & Webster, E. 2000. Assessing long-range transport potential of persistent organic pollutants. *Environmental Science & Technology*, 34(4), 699–703.

Boholm, Å. 1996. Risk perception and social anthropology: Critique of cultural theory. *Ethnos*, 61(1–2), 64–84.

Boyd, E., Cornforth, R.J., Lamb, P.J., Tarhule, A., Lélé, M.I., & Brouder, A. 2013. Building resilience to face recurring environmental crisis in African Sahel. *Nature Climate Change*, 3(7), 631–637.

Brown, P., Zavestoski, S.M., & McCormick, S., Mandelbaum, J., & Luebke, T. 2001. Print media coverage of environmental causation of breast cancer. *Sociology of Health & Illness*, 23(6), 747–775.

Burger, E.J. 1990. Health as a surrogate for the environment. *Daedalus*, 119(4), 133–153.

Burton, I., Kates, R.W., & White, G.F. 1993. *The Environmental as Hazard*. New York: Guilford.

Corbett, J.B. 2001. Women, scientists, agitators: Magazine portrayal of Rachel Carson and Theo Colborn. *Journal of Communication*, 51(4), 720–749.

Dean, M. 1999. 'Risk, Calculable and Incalculable' in Lupton, D. (ed.). *Risk and Socio-cultural Theory: New Directions and Perspectives*. Cambridge: Cambridge University Press. pp 131–159.

Douglas, M. 1966. Purity and Danger: *An Analysis of Concepts of Pollution and Taboo.* New York: Praeger.

Douglas, M., & Wildavsky, A. 1982. How can we know the risks we face? Why risk selection is a social process. *Risk Analysis*, 2(2), 49–58.

Douglas, M., & Wildavsky, A. 1983. *Risk and Culture: An Essay on the Selection of Technological and Environmental Dangers*. Berkeley: University of California Press.

Durnford, D., Dastoor, A., Figueras-Nieto, D., & Ryjkov, A. 2010. Long range transport of mercury to the Arctic and across Canada. *Atmospheric Chemistry and Physics*, 10(13), 6063–6086.

Environment Canada. 2014a. *What is the Great Lakes Water Quality Agreement?* Available at: https://www.ec.gc.ca/grandslacs-greatlakes/default.asp (Accessed 7 January 2015).

Environment Canada. 2014b. *History of the Great Lakes Water Quality Agreement.* Available at: https://www.ec.gc.ca/grandslacs-greatlakes/default.asp (Accessed 7 January 2015).

Eyles, J., & Fried, J. 2012. 'Technical breaches' and 'eroding margins of safety' – Rhetoric and reality of the nuclear industry in Canada. *Risk Management*, 14(2), 126–151.

Fluoride Action Network. 2014. *Communities Which Have Rejected Fluoride Since 1990.* Available at: http://fluoridealert.org/content/communities/ (Accessed 7 January 2015).

Foucault, M., Burchell, G., Gordon, C., & Miller, P. (Eds.). 1991. *The Foucault Effect: Studies in Governmentality*. Chicago: University of Chicago Press.

Funtowicz, S.O., & Ravetz, J.R. 1993. Science for the post-normal age. *Futures*, 25(7), 739–755.

Furimsky, E. 2002. Sydney tar ponds: Some problems in quantifying toxic waste. *Environmental Management*, 30(6), 872–879.

Gibbs, L.M. 1995. *Dying from Dioxin: A Citizen's Guide to Reclaiming Our Health and Rebuilding Democracy*. Boston: South End Press.

Giddens, A. 1990. *The Consequences of Modernity*. Cambridge, UK: Polity.

Giddens, A. 1991. *Modernity and Self-Identity: Self and Society in the Late Modern Age*. Cambridge: Polity.

Giddens, A. 1999. *Runaway World: How Globalization Is Reshaping Our Lives*. London: Profile.

Heath, C.W., Nadel, M.R., Zack, M.M., Chen, A.T., Bender, M.A., & Preston, R.J. 1984. Cytogenetic findings in persons living near the Love Canal. *JAMA*, 251(11), 1437–1440.

Hewitt, K. (Ed.). 1983. *Interpretations of Calamity from the Viewpoint of Human Ecology* (No. 1). Winchester, MA: Allen & Unwin.

Intergovernmental Panel on Climate Change. 2015. *2014 Synthesis Report (5th Assessment)*. Available at: http://www.ipcc.ch/report/ar5/syr/ (Accessed 7 January 2015).

International Joint Commission. 2014. *Role of the IJC.* Available at: http://www.ijc.org/en_/Role_of_the_Commission (Accessed 7 January 2015).

Jaffe, D., Anderson, T., Covert, D., Kotchenruther, R., Trost, B., Danielson, J., & Uno, I. 1999. Transport of Asian air pollution to North America. *Geophysical Research Letters*, 26(6), 711–714.

Jasanoff, S. 2002. Citizens at risk: Cultures of modernity in the US and EU. *Science as Culture*, 11(3), 363–380.

Kasperson, R.E., Renn, O., Slovic, P., Brown, H.S., Emel, J., Goble, R., & Ratick, S. 1988. The social amplification of risk: A conceptual framework. *Risk Analysis*, 8(2), 177–187.
Kitzinger, J. 1999. Researching risk and the media. *Health, Risk & Society*, 1(1), 55–69.
Knutson, T.R., McBride, J.L., Chan, J., Emanuel, K., Holland, G., Landsea, C., . . . & Sugi, M. 2010. Tropical cyclones and climate change. *Nature Geoscience*, 3(3), 157–163.
Kunreuther, H., Easterling, D., Desvousges, W., & Slovic, P. 1990. Public attitudes toward siting a high-level nuclear waste repository in Nevada. *Risk Analysis*, 10(4), 469–484.
Latour, B., & Woolgar, S. 2013. *Laboratory Life: The Construction of Scientific Facts*. Princeton, NJ: Princeton University Press.
Lebedynets, M., Sprynskyy, M., Kowalkowski, T., & Buszewski, B. 2005. Evaluation of hydrosphere state of the Dniester river catchment. *Polish Journal of Environmental Studies*, 14(1), 65–71.
Lerer, L.B., & Scudder, T. 1999. Health impacts of large dams. *Environmental Impact Assessment Review*, 19(2), 113–123.
Liang, Q., Jaeglé, L., Jaffe, D.A., Weiss-Penzias, P., Heckman, A., & Snow, J.A. 2004. Long-range transport of Asian pollution to the northeast Pacific: Seasonal variations and transport pathways of carbon monoxide. *Journal of Geophysical Research: Atmospheres (1984–2012)*, 109(D23), 1–9.
Luhmann, N. 1989. *Ecological Communication*. Chicago: University of Chicago Press.
Luhmann, N. 1993. *Communication and Social Order: Risk: A Sociological Theory*. New York: A. de Gruyter.
Lupton, D. 1999. *Risk and Sociocultural Theory: New Directions and Perspectives*. Cambridge: Cambridge University Press.
Mitka, M. 2012. Rigorous evidence slim for determining health risks from natural gas fracking. *Journal of the American Medical Association*, 307(20), 2135–2136.
Murdoch, J.C., & Sandler, T. 1997. The voluntary provision of a pure public good: The case of reduced CFC emissions and the Montreal Protocol. *Journal of Public Economics*, 63(3), 331–349.
Mythen, G. 2004. *Ulrich Beck: A Critical Introduction to the Risk Society*. Sterling, VA: Pluto Pr.
Neuman, W.R. 1990. The threshold of public attention. *Public Opinion Quarterly*, 54(2), 159–176.
Perrow, C. 2011. *Normal Accidents: Living with High Risk Technologies*. Princeton, NJ: Princeton University Press.
Pielke Jr, R.A., Landsea, C., Mayfield, M., Laver, J., & Pasch, R. 2005. Hurricanes and global warming. *Bulletin of the American Meteorological Society*, 86(11), 1571–1575.
Polizzotto, M.L., Kocar, B.D., Benner, S.G., Sampson, M., & Fendorf, S. 2008. Near-surface wetland sediments as a source of arsenic release to ground water in Asia. *Nature*, 454(7203), 505–508.
Rabe, B.G. 1994. *Beyond NIMBY: Hazardous Waste Siting in Canada and the United States*. Washington, DC: Brookings Institution.
Reich, M.R. 1983. Environmental politics and science: The case of PBB contamination in Michigan. *American Journal of Public Health*, 73(3), 302–313.
Renn, O. 2008. *Risk Governance: Coping with Uncertainty in a Complex World*. Sterling, VA: Earthscan.
Richter, I., Berking, S., & Müller-Schmid, R. (Eds.). 2006. *Risk Society and the Culture of Precaution*. New York: Palgrave Macmillan.
Rosa, E., McCright, A., & Renn, O. 2014. *The Risk Society Revisited: Social Theory and Risk Governance*. Philadelphia, PA: Temple University Press.

Schiermeier, Q. 2012. Hot air: Commitments made under the Kyoto climate treaty will expire at the end of 2012, but emissions are rising faster than ever. *Nature*, 491(7426), 656–658.

Shen, L., Wania, F., Lei, Y.D., Teixeira, C., Muir, D.C., & Bidleman, T.F. 2005. Atmospheric distribution and long-range transport behavior of organochlorine pesticides in North America. *Environmental Science & Technology*, 39(2), 409–420.

Shepherd, D., McBride, D., Welch, D., Dirks, K.N., & Hill, E.M. 2011. Evaluating the impact of wind turbine noise on health-related quality of life. *Noise and Health*, 13(54), 333.

Slovic, P., Flynn, J., & Gregory, R. 1994. Stigma happens: Social problems in the siting of nuclear waste facilities. *Risk Analysis*, 14(5), 773–777.

Tansey, J., & Rayner, S. 2009. 'Cultural Theory and Risk' in Heath, R.L., & O'Hair, D.H. (eds.). *Handbook of Risk and Crisis Communication*. London: Routledge. pp 53–79.

Thomas, W.I. & Thomas, D.S. 1928. *The Child in America: Behavior Problems and Programs*. New York: Knopf.

United Nations. 2014. *United Nations Conference on Environment & Development*. Available at: http://sustainabledevelopment.un.org/content/documents/Agenda21.pdf (Accessed 7 January 2015).

United Nations Economic Commission of Europe. 2014. *Convention on Long-Range Transboundary Air Pollution*. Available at: http://www.unece.org/env/lrtap (Accessed 7 January 2015).

United Nations Environment Program. 2014. *The Montreal Protocol on Substances that Deplete the Ozone Layer*. Available at: http://ozone.unep.org/new_site/en/montreal_protocol.php (Accessed 7 January 2015).

United Nations Framework Convention on Climate Change. 2014. *Kyoto Protocol*. Available at: http://unfccc.int/kyoto_protocol/items/2830.php (Accessed 7 January 2015).

Velders, G.J., Andersen, S.O., Daniel, J.S., Fahey, D.W., & McFarland, M. 2007. The importance of the Montreal Protocol in protecting climate. *Proceedings of the National Academy of Sciences*, 104(12), 4814–4819.

Victor, D.G. 2004. *The Collapse of the Kyoto Protocol and the Struggle to Slow Global Warming*. Princeton, NJ: Princeton University Press.

Vorkamp, K., & Rigét, F.F. 2014. A review of new and current-use contaminants in the Arctic environment: Evidence of long-range transport and indications of bioaccumulation. *Chemosphere*, 111, 379–395.

Vyner, H.M. 1988. *Invisible Trauma: The Psychosocial Effects of Invisible Environmental Contaminants*. Lexington, MA: Lexington Books/DC Heath and Com.

Wallenius, M.A. 2004. The interaction of noise stress and personal project stress on subjective health. *Journal of Environmental Psychology*, 24(2), 167–177.

Webster, P.J., Holland, G.J., Curry, J.A., & Chang, H.R. 2005. Changes in tropical cyclone number, duration, and intensity in a warming environment. *Science*, 309(5742), 1844–1846.

WHO. 2014. *Global Health Observatory: Life Expectancy*. Available at: http://www.who.int/gho/mortality_burden_disease/life_tables/situation_trends_text/en/ (Accessed 7 January 2015).

World Nuclear Association. 2014. *Storage and Disposal Options: Radioactive Waste Management Appendix 2*. Available at: http://www.world-nuclear.org/info/nuclear-fuel-cycle/nuclear-wastes/appendices/radioactive-waste-management-appendix-2—storage-and-disposal-options/ (Accessed 7 January 2015).

Wynne, B. 1992. Uncertainty and environmental learning: Reconceiving science and policy in the preventive paradigm. *Global Environmental Change*, 2(2), 111–127.

Zinn, J. (Ed.). 2008. *Social Theories of Risk and Uncertainty: An Introduction*. Malden, MA: Blackwell.

6 Is the risk reversible?

Introduction

We live largely in a world of fear where risk has powerful negative connotations and consequences. This fear seems to occur irrespectively of geographical scale. The World Economic Forum (2014) highlights the global risk landscape, linking all global risks together (see Figure 6.1). We should note that health is very much seen as a consequence of these risks. They are contained within the societal class, relating to social stability (e.g., income disparities, food crises) and public health (e.g., pandemics, antibiotic-resistant microorganisms, chronic disease burden) dimensions. This category has decreased in relative importance since 2010. Yet a healthy population is a necessary condition for sustained economic development.

The WHO (2014) in its comments on macro-economic policy has noted that good health matters for economic growth through higher labour productivity, demographic changes and higher educational attainment. Poor health undermines economic growth, so Africa's per capita growth rate of 0.4% in 1990–1997 was three times lower than it would have been had HIV/AIDS not existed. In terms of crude indicators, increased life expectancy can result in higher levels of economic growth – one year of increased life expectancy in the population leads to an average 4% rise in economic growth (Bloom et al., 2004). Better health can also promote higher real net income. This is not a straightforward relationship and other social, and environmental forces may militate against it.

With respect to the Millennium Development Goals, there has been lack of progress overall but especially with respect to environmental matters, where on some indicators there has been a reversal and deterioration. WHO (2006, p. 9) estimated that

> 24% of the disease burden (healthy life years lost) and an estimated 23% of all deaths (premature mortality) was attributable to environmental factors. Among children 0–14 years of age, the proportion of deaths attributed to the environment was as high as 36%. There were large regional differences in the environmental contribution to various disease conditions – due to differences in environmental exposures and access to health care across the regions. For example, although 25% of all deaths in developing regions were attributable

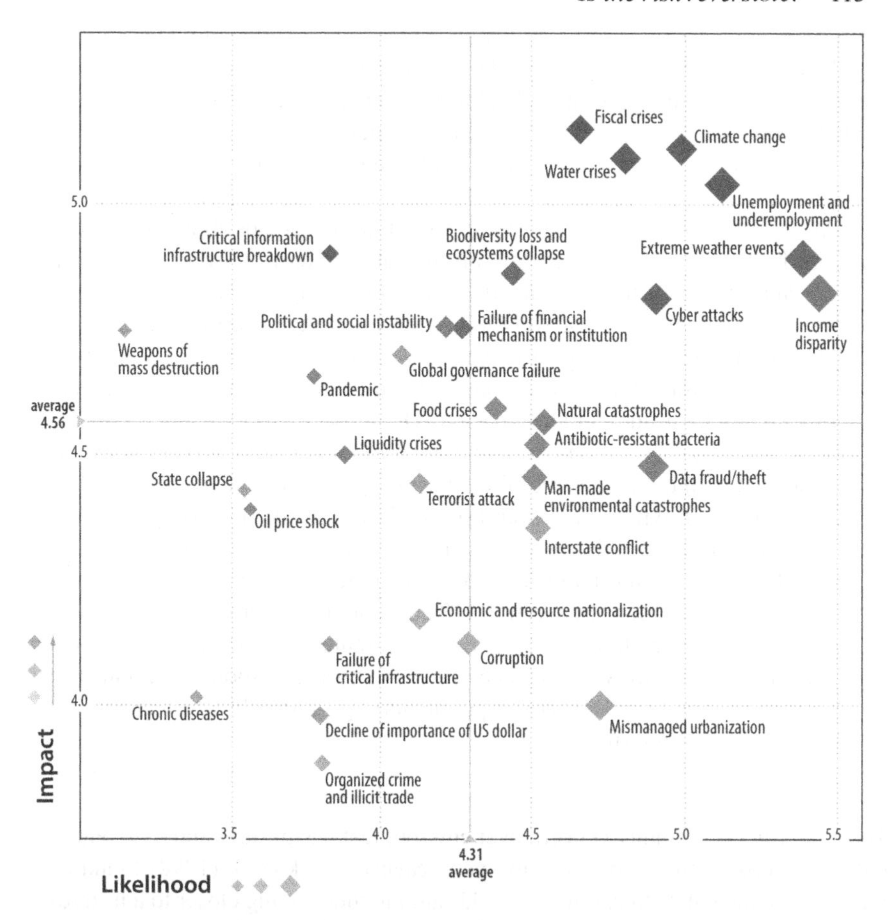

Figure 6.1 Global risks landscape

Source: World Economic Forum (2014), with kind permission of the World Economic Forum. Please see original report (http://reports.weforum.org/global-risks-2014/) to view color. Key – blue: economic; green: environmental; orange: geopolitical; red: societal; purple: technological

to environmental causes, only 17% of deaths were attributed to such causes in developed regions.

Such environmental risks seem to be largely a matter of low- and middle-income countries. Yet this is not the case, especially when we consider not human neglect but conscious human activity.

Relationships between environmental exposure and poor health are documented to a certain extent. For example, the U.S. EPA (2014) in its report on the environment notes that some cancers, heart disease, respiratory outcomes, infectious diseases and early infant challenges have been associated with a wide array of exposures (see Chapter 1). But it comments that definitive conclusions are difficult to assess, as the U.S. population is very diverse and most diseases are multi-factorial in terms of associations. In epidemiologic terms of understanding population health, these are 'confounders' of the relationship between, for example, chemical exposures and poor health. But some remain and continue to be publicly worrying, especially those relating to childhood and developmental matters. The report summarizes that state of evidence for birth outcomes as follows:

> Environmental exposures are being investigated for possible associations with birth outcomes such as low birthweight, preterm delivery, and infant mortality. Some of the risk factors for low birthweight infants born at term include maternal smoking, weight at conception, and nutrition and weight gain during pregnancy. Specific examples of known or suspected environmental contaminant influences on birth outcomes include environmental tobacco smoke, lead, and air pollution. The most robust evidence exists for environmental tobacco smoke and lead. Environmental tobacco smoke is associated with increased risk of low birthweight, preterm delivery, and sudden infant death syndrome. Several studies have identified lead exposure as a risk factor for preterm delivery. Associations between air pollution and foetal growth and infant mortality have been documented. Recent studies report significant associations between PM10 concentration averaged over a month or a trimester of gestation and the risk of intrauterine growth reduction and low birthweight. Growing evidence shows exposure–response relationships between maternal exposures to air pollutants (e.g., sulfur dioxide and particulates) and preterm birth.
>
> EPA (2014)

There is strong community concern over environmental exposures and their potential for adverse birth outcomes. With many caveats, Dolk et al. (1998) found an increased chance of birth anomalies (1.33) among those living closer to a hazardous waste facility than the control populations, after adjusting for socio-economic status and maternal age. Among congenital anomalies, neural tube defects (1.86) and malformations of the cardiac septa (1.49) were significantly over-represented among those living within 3 km of a hazardous site. Neural tube defects were a cause celebre in Brownsville, Texas, where in April 1991 three anencephalic babies had been delivered in one hospital within 36 hours. Active surveillance revealed that, in fact, at this hospital, six babies with neural tube defects (NTDs) had been delivered within 6 weeks. For a county with 6,000 births each year, two or fewer cases would have been expected in a year. At the time, people living in the border communities expressed concern that the NTDs were caused by pollution, with agricultural pollution, industrial pollution from the unregulated industries (maquiladoras) immediately across the border, and the consumption of contaminated fish caught from nearby rivers being implicated. There was also potential for pesticide exposure as well as the characteristics of the predominantly Mexican American population (93%) – high rates of poverty and adverse health conditions such as obesity and diabetes; a corn-based diet; substandard housing; and use of folk medicines and unique access to over-the-counter prescription medications in Mexican pharmacies. Many studies were undertaken and focused not only on environment but also diet, nitrate exposure in drinking water and nitrites in over-the-counter drugs. The studies found a complex system of association involving NTD incidence, with Suarez et al. (2012) concluding that a population deficient in folic acid will be vulnerable to a variety of insults whether brought on by individual behaviours (e.g., obesity) or through the surrounding environment (e.g., fumonisins – molds). Regardless of the cause in this particular case, the extent of study of the problem highlights that public concern over children and their health and well-being reflects a concern for the future of humankind. The is true of a series of studies attempting to tease out the link between a skewed gender ratio and high levels of local industrial pollution in the Aamjiwnaang First Nation near Sarnia, ON (Mackenzie et al., 2005; Luginaah et al., 2010). The studies were spawned when local parents noticed there were not enough boys to form local minor-league sports teams. From 1994 through 2003, the time of the study, the number of male births significantly declined to a 10-year ratio of 0.41 and a 5-year ratio of 0.35. A number of environmental, chemical and occupational exposures are implicated given the large nearby petrochemicals industries, and Van Larebeke et al. (2008) in particular single out the biologically plausible mechanism of endocrine disruption whereby environmental chemicals interfere with normal hormone activity in the body.

Brain tumours on the other hand reflect a concern about effective functioning in society. The identification of 12 brain tumour deaths by a community resident around Los Alamos in 1991 caused great public concern. The alarm received local and national press coverage and investigations by the New Mexico Health Department. These investigations examined the period 1970 to 1990 and found elevated rates in the period 1986 to 1989 with 10 reported cases (3.9 expected). Yet later academic work found that size of study area, population size and number of cases affected the 'cluster', leading them to conclude that the excess of brain cancer falls within the realms of chance (see Kulldorff et al., 1998). But later work under New Mexico's Right to Know found brain cancer rates elevated, if not statistically significant, and other cancers, including breast, ovary, prostate and thyroid to be statistically significantly elevated (see Concerned Citizens for Nuclear Safety, 2003; Price, 2011). But considering several cancer sites brings potentially other associations to the fore. Who benefits from this segregation of diseases? And does it make the problems reversible with better technical protections or lifestyle changes?

Lifestyle changes

It is commonplace to argue that individuals and communities can minimize and perhaps reverse the adverse consequences of pollution via simple lifestyle changes. Thus the EPA (2010) describes non–point-source pollution as people pollution created by daily activities such as using cleaners, detergents and pesticides carefully or not at all and by the careful disposal of garbage to ensure clean water and soil. Yet lifestyle is also a 'confounder' in environment and health studies in that a wide array of activities in which people engage can have positive and negative impacts on health outcomes of concern in environment and health investigations – including diet, exercise and sexual behaviours, but also use of various landscapes and spaces.

There is further a broader argument that non-communicable diseases (such as diabetes, cardiovascular disease and cancer) are clearly associated with modifiable health risk behaviours, such as physical inactivity and improper food consumption. Many individual and societal factors are involved. North Americans and Europeans fail to engage in recommended levels of physical activity (30–60 minutes of moderate intensity on most days of the week) and proper dietary behaviours (e.g., eating 5 to 10 fruits and vegetables a day, limiting total fat intake to 20 to 35% of daily intake; see Sweet and Fortier, 2010). It seems that only 8 to 10% of the population participate in a sufficient amount of physical activity and follow dietary recommendations in the U.S. Similar proportions have been found in the UK and the Netherlands. However, these rates are low given that the benefits of participating in the recommended levels of physical activity and adhering to dietary recommendations are well established. Later, we examine why this irreversibility of poor health behaviours has not occurred. Individual activities can help, but "the combined effects of all diet and physical activity related behaviours that affect health directly and via their effects on obesity have a much larger total effect than any separate pathway" (Popkin et al., 2006, p. 289). For most people, adherence to behavioural change interventions is difficult, and there are also the systemic factors that shape behaviour generally. The role of government subsidies on some food products, the design of streetscape and city environments, the political power of Big Food and the restrictive practices of trade agreements may also conspire against individual changes.

There are also individual limitations to change. Some 15 years ago, Stern (2000) commented on letting faith in people's actions dominate evidence about these actions. He notes that the relative importance of individual as opposed to organizational activity must be considered, with the former more important for some (e.g., those associated with vehicle use) and the latter for others (solid waste production). Furthermore, while individuals consume, much consumption is determined by organizational decisions about the production of goods and services as well as the need for accessibility to necessary goods and services. There also seems to be little desire to sacrifice the benefits of technology, even if less pollution or contamination is produced. It is a slow process to change individuals' behaviours so that they do such things as use public transit, buy energy-efficient cars and use the latest home energy-efficiency technologies like heat exchangers. These new technologies may also not produce the benefits their advocates suggest. Furthermore, ideas and interests intercede to prevent many individuals and communities from adopting pro-environmental attitudes and turning those into actions. These tie in with the cultural worldviews described in Chapter 2. There are also strong situational pressures which may prevent a pro-environmental response to a policy suggestion or an educational program. These pressures, such as lack of nearby facilities, cost of alternative actions and behavioural inertia from the relative ease of pursuing routinized patterns, even if motivation has developed the idea that individual action may not lead to significant change, can result in doing nothing and/or seeing the problem as overwhelming, leading to it being ignored.

What Stern identifies is a collective action problem in which a decision is made not to participate in pro-environmental decisions, even for health benefit. Rydin and Pennington (2000) have discussed this with respect to participation in environmental planning in which non-co-operative behaviour (such as shirking and free-riding) may impact the effectiveness of the process. Public choice theory argues that an individual's efforts will have an impact less than the cost of her/his participation. The outcomes of participation are non-exclusionary and indivisible, so the rational strategy for the individual is to be a free rider gaining the benefits of others' involvement at no personal cost. Even if many individuals do, say, change their behaviours, their participation is often unstable and unsustainable. Furthermore, if the individual direct benefits are small or discounted in the future, it is unlikely that individuals will even want to become informed about the situation itself – bike lanes or transit-only lanes might be examples which then have to be regulated into existence.

That said, even within rational economic theory there are signs of hope in the conundrum that people often donate to public goods despite the fact that the donating individual is not likely to receive any direct benefit. The concept often used to explain this is the 'warm glow of giving' and that people 'do the right thing' because it makes them feel good, and this does not apply simply to the wealthy who presumably can afford to. That is, willingness to pay for enhanced environmental assets and willingness to accept payment for environmental degradation studies show that there is an asymmetry in favour or willingness to pay for good things than accept compensatory payments for harm (Kahneman and Knetsch, 1992; Andreoni, 1995). The trick then is to tap into willingness to pay hypothetical dollar amounts into behavioural changes that take time (e.g., waste separation) or wholesale lifestyle changes (e.g., moving to higher-density urban dwellings that maximize household and city-wide energy efficiencies).

With respect to public health, Siegal et al. (2009, p. 1584) have argued that

> public goods are created or maintained as the result of successful concerted action. They are by definition nonexcludable and are typically nonrivalrous. This means that everyone can enjoy the benefits of a public good (e.g., herd immunity, clean parks, clean air, a standing army for national defense) even if some individuals do not contribute or serve their personal interests at the expense of the common interest (e.g., littering or polluting). Yet if enough people fail to contribute to the public good or engage in counterproductive acts, the public good is threatened.

Environmental degradation or pollution can thus cause a tragedy of the commons wherein the quality of collective goods is eroded through, for example, poor air quality, inadequate water quality and quantity. Furthermore, aggregate decision making to maintain public goods is prone to herding (individuals will pursue the consensus and refrain from making deviant yet positive environmental choices), groupthink (groups strive for unanimity even at the expense of quality decision making) or social loafing (in groups, people tend to feel unmotivated because they consider their contributions unnoticeable or not evaluated). Siegal et al. discuss this in the context of organ donation where the very rules – opt-in as opposed to opt-out – feed risk aversion, the overestimation of costs in terms of benefits and inertia, all of these having parallels in environmental change like CO_2 emission reductions at multiple scales (household to nation states).

For changing behaviours at the local level, national efforts have begun to use other ideas from behavioural economics, namely those connected to nudging. A nudge is an attempt to change people's behaviour to make more rational decisions (as defined by experts) without limiting their options. There are claims that a nudge is a form of a cost, psychological as opposed to economic, and that there are manipulative, changing preferences and privileging long-term over short-term utilities (see Fischer and Lotz, 2014). For example, Quigley (2013) argues that providing community bicycles might nudge individuals to reduce car use. Yet nudging is widely practised mainly through policy defaults through which people are encouraged in terms of risk and cost to maintain the default option. Opting out is made difficult. Several governments have established behavioural insight units to alter eating behaviour, energy consumption and so on. But as in all attempts to alter preferences, context and its social values are important. Simple changes of language and signs can change behaviour if they fit with social norms. In Denmark, one trial put green arrows pointing to the stairs next to railway-station escalators in the hope of encouraging people to take the healthier option. This had almost no effect. The other trial had a series of green footprints leading to garbage containers. These signs reduced littering by 46% during a controlled experiment in which wrapped sweets were handed out. There appear to be "no social norms about taking the stairs but there are about littering" (see "Nudge Nudge, Think Think," 2012a). This may link back to the warm glow idea vis-à-vis society and individuals: in that a litter-free environment is a visible collective benefit while taking the stairs may be seen more as an individual health benefit. In the UK, LGA (2013) notes that nudging is only one type on a continuum of interventions to alter behaviour, along with smacks (banning goods and services), shoves (financial disincentives – tobacco taxes) or nudges (information provision, altering building design or changing the default). Examples of nudges with respect to obesity and exercise include making salad not fries the default or providing a collective incentive gained by individual participation in the activity. Caversham, Reading, UK, was the locale of a challenge to 'Beat the Street' by residents walking the distance of twice round the world to gain money for local and school library books.

Yet many environmental exposures and impacts have an aggregate effect on health and require more smacking and shoving as well as the use of a strong evidence base. In these instances, firms may actively work against policy and behavioural change, often unless or until there is a technological solution to the environmental concern – health matters little. Lead in gasoline is a case in point (see Chapter 3). Other examples include ozone layer depletion and the use of bisphenol A (BPA) in food containers. At the time that Health Canada (2014) invoked the ALARA (as low as reasonably achievable) principle regarding bisphenol A in infant feeding bottles, there were already non–BPA alternatives available on the marketplace, so relatively perfect substitution was possible.

The switch away from chlorofluorocarbons (CFCs) has been far slower. CFCs are mainly responsible for human-induced ozone depletion. Ozone is responsible for determining the amount of UVB radiation on earth, and the public health concerns are for increases in skin cancer and cataracts. This relationship between CFCs (used in air conditioning and cooling units, as aerosol spray propellants and in the cleaning processes of delicate electronic equipment) and ozone depletion was first hypothesized in 1974. The aerosol and halocarbon industries were highly critical, but by the mid-1970s a scientific consensus emerged which led to the banning of CFCs in aerosol sprays. International action followed, and in 1987 43 countries signed on to the Montreal Protocol to phase out CFCs (see Rowland, 2007). Their actions were further promoted by the identification of a growing ozone hole over Antarctica, with public concerns over radiation exposure and other invasions from space. Industry changed its mind, and progress towards targets has been slow. CFCs can be replaced by a hydrocarbon refrigerant developed in 1992 but are still used by less than half of the refrigerator market and contested by industry especially in the U.S. Other users have changed to hydrochlorofluorocarbons (HCFCs), which contain no chlorine or bromine and do not contribute to ozone depletion, although they are potent greenhouse gases. These science-based policy shifts have reduced the rate of ozone depletion and the ozone hole. There are however chemicals not regulated by the Protocol – nitrous oxide, produced in sewage treatment, combustion and on factory farms – which have replaced CFCs as the largest ozone-depleting and greenhouse-producing gas to the atmosphere (see Ravishankara et al., 2009). Reversal of risk may lead to new risks emerging.

Though BPA has been in use for decades, action was fairly swift as evidence mounted and despite the politics of industry pushback. BPA is a carbon-based synthetic compound which is a colourless solid that is soluble in organic solvents but poorly soluble in water. It has been in commercial use since 1957. BPA is used to make certain plastics and epoxy resins. BPA–based plastic is clear and tough, and it is present in common consumer goods, such as water bottles, sports equipment, CDs and DVDs. Epoxy resins containing BPA are used to line water pipes, as coatings on the inside of many food and beverage cans and in making thermal paper used in sales receipts. BPA exhibits hormone-like properties that raise concern about its suitability in some consumer products and food containers. Since 2008, several governments have investigated its safety with bans on its use in baby bottles in Canada and the European Union. This prompted some retailers to withdraw polycarbonate products. The U.S. FDA has ended its authorization of the use of BPA in baby bottles and infant formula packaging, based on market abandonment, meaning that no judgement on safety is required. BPA is ubiquitous. As Walsh (2010) put it,

> the problem is, BPA is . . . a synthetic estrogen, and plastics with BPA can break down, especially when they're washed, heated or stressed, allowing the

> chemical to leach into food and water and then enter the human body. That happens to nearly all of us; the CDC has found BPA in the urine of 93% of surveyed Americans over the age of 6. If you don't have BPA in your body, you're not living in the modern world.
>
> (*Time* online http://content.time.com/time/specials/packages/article/0,28804,1976909_1976908_1976938-2,00.html)

While the plastics industry fought the ban on BPA in baby bottles, they have accepted the rule of government and the power of large retailers (e.g., Toys R Us, Walmart), which withdrew BPA–based products. But the economic impacts on the industry were small, and manufacturers of plastic bottles were quick to find alternatives. In 2008 Nalgene announced that they would phase out voluntarily the use of polycarbonate containing BPA. They manufacture consumer bottles and containers using a variety of non–BPA plastics. Playtex, another large supplier, has done the same.

Bringing us back to the theme of child health being a flashpoint for environment and health study and action, BPA in baby bottles and sippy cups was framed as an attack on children and therefore humanity's future. This exposure was removed. Yet other plastics remain a potential threat to human health. For example, phthalates are mainly used as plasticizers (substances added to plastics to increase their flexibility, transparency, durability and longevity). Phthalates are easily released to the environment, and as they degrade rapidly, interior exposures are of greatest potential damage. They have been removed from children's toys in most jurisdictions, but there are as yet no convenient and inexpensive alternatives.

"When science meets values head on"

BPA provides an example of interests and ideologies that underpin the use of scientific evidence. Canada was the world leader in banning BPA in baby bottles. Edge and Eyles (2013) note that in April 2008, Health Canada released a risk assessment of BPA, concluding that the general public need not be concerned, as estimated exposure levels fell below regulatory safety standards, and adverse health effects were not expected. However, they felt the margin of safety was too small for infant exposures and thus warranted a precautionary response. Consequently, in March 2010, the government of Canada moved to formally prohibit the advertisement, sale and importation of polycarbonate plastic baby bottles containing bisphenol A to reduce infant exposure. Additionally, in October 2010, the government formally categorized the substance as 'toxic' under the Canadian Environmental Protection Act, despite objections from industry, some scientists and policy makers in other jurisdictions that the current weight of evidence did not justify these precautionary measures. Why? Science is not of course everything, and its uses and interpretations are in many ways culturally dependent, shaping how arguments are legitimized as credible (or not) within governance and regulatory processes. Even a precautionary approach must fit with a country's economic and international trade obligations. There were also legal and political contexts.

Timing, state of the economy and political dynamics within and between political parties and other interests were important. In 1999, it was ruled that the 23,000 substances already in commercial use that had never undergone risk assessment must be categorised within a seven-year time limit for review, the clock starting in 1999. This time expired with a Conservative minority government in power. By the end of 2006 and prior to the economic downturn, Canadian public opinion polls identified the environment as the most important issue, with a fivefold increase in support unfolding over a year (Edge and Eyles, 2013). The needs of political leaders cannot be ignored, with the Canadian prime minister, perceived as economically liberal and socially conservative, being the person to announce the 'Plan' in 2006 and its tight public health protective timeline of three years. The Plan allowed regulators to find substances toxic unless proven otherwise. This was the case with bottles and cups but not cans – yet the government could claim to be protective of the nation's children.

Science meets politics on the global stage: The case of climate change

Political interests, scientific evidence and technological alternatives appear to have led to a reversal in identified risks. This is not the case if the problem appears so integrated with the ways in which society operates – the messiest collective action problem is likely to be climate change.

Despite widespread scientific agreement about anthropogenic climate change, Estry and Moffa (2012, p. 777) begin their paper:

> Climate change stands out as the quintessential global-scale collective action problem with implications that require carefully managed policy coordination and multi-level governance. The build-up of greenhouse gases in the atmosphere threatening global warming, sea level rise, changed rainfall patterns (leading to shifting agricultural productivity), and increased intensity of hurricanes and typhoons can neither be successfully addressed by any one nation acting alone nor by governmental action at any one level of geographic scale.

The problems of free-riding and competitive disadvantage from independent action are significant. Sources of past and present greenhouse gas emissions are different, as are the costs and benefits of undertaking no mitigating actions. And while there is agreement that international co-operation is needed there is conflict over the rules for burden-sharing, especially when for most emitting countries the costs, especially for human health, are in the future, thus discounting the problems to later generations. There is also a complex regulatory system with overlapping international agencies, often challenged by the self-interests of national governments. Furthermore, climate change is seen as a complex issue, with many publics seeing jobs and economic development, which may lead to greater energy use and higher levels of greenhouse gas emissions. The environment seems to be a public concern during times of economic prosperity. For example, Gallup added climate change (see Figure 6.2) to its list of concerns that worry Americans for

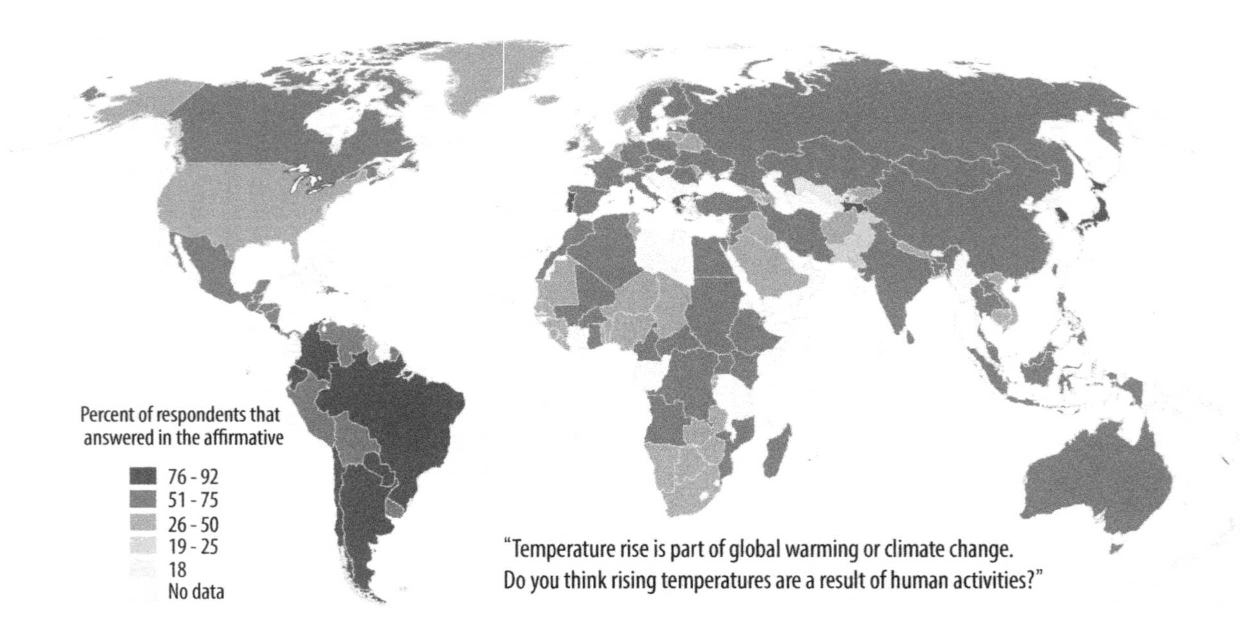

Figure 6.2 Opinions that climate change is anthropogenic by country 2008–2009

Source: GunnMap (2010), under CC BY-SA 4.0.

the first time in 2014. More than half of respondents considered it to be of little or no concern (Riffkin, 2014). Thirteen other worries are seen as more troubling. Gallup asked about global warming in earlier surveys, which note its decline from two-thirds to a half between 2007 and 2011 of those who are concerned (Gillis, 2012). This changing public uncertainty is used by interest groups with specific value sets to reduce the impact and even development of mitigation measures. In Australia, for example, Hamilton (2007) notes how the fossil-fuel industry, particularly coal exporters, have worked assiduously against regulatory change. Michaels (2008) shows how the operation of some industrial lobby groups with respect to climate change is similar to those who advocated for tobacco companies and chemical and plastics companies (lead, BPA, asbestos), often supported by parts of government agendas (e.g., Canadian support for the asbestos industry in Quebec, oil sands in Alberta). For how long can special interests and an uninterested publics challenge the scientific consensus about climate change? Are health risks from climate change not reversible or does the time scale and multi-factor associations of environment and disease make the question unanswerable without much remaining uncertainty? We leave this discussion to a later chapter.

Technological changes

The reversibility of risk, especially from collective activities, is often seen as a technological fix. It is impossible to overestimate the benefits which the technological improvements in air and water quality, sewage disposal and changes in construction codes for buildings and food safety brought to public health for many populations. The importance of housing as an environmental and public health issue has been recognized since the 19th century with the commentaries by Engels on the conditions of the working class in England. Chadwick and Riis documented the unsanitary conditions that resulted in a massive movement to change conditions (see Krieger and Higgins, 2002). Sanitary reform is said to have greater impact on improving health than medical interventions. Rosen (1958) reports the targets of the reformers. "They educated the public on hygiene, lobbied for policy reform, and sought to eliminate crowded, poorly ventilated, and filthy [housing], impure water supplies, inadequate sewerage, and unwholesome food". He goes on that in New York City, the Council of Hygiene's report on the sanitary conditions of the city resulted in the country's first health and housing laws. What followed was legislation requiring windows that opened to outside air in place of air shafts, separate 'water closets' for each apartment, functional fire escapes, adequate lighting in hallways, proper sewage connections and regular waste removal. Such acts reduced the development and severity of infectious disease outbreaks. Other challenges have arisen such as the use of lead in paint and gasoline and its impact on children in particular. Further attention has been given to the indoor environment, where cold, mouldy damp and dusty conditions can impact respiratory health (see Dales et al., 1991).

These issues remain challenges in low-income settings and countries. Poor-quality ventilation, source of energy used, lack of screening, poor construction, intermittent electricity supply and poor water quantity and quality have all been identified as health concerns which have a technological solution (see Haines et al., 2012, p. 825), suggesting that

> the rapidly increasing number of houses built in developing countries both in rural and urban areas offer real opportunities for improving health by incorporating easily installed, affordable features such as screens on doors, windows and ceilings and considering the most appropriate and affordable ways to provide water and sanitation.

Yet demand is massive and funding low. For example in South Africa, the 2011 census shows

> that households that have flush toilets connected to the sewerage system increased to 57% in 2011 from 50% recorded in Census 2001. Households with no toilets declined significantly from the 13,3% in 2001 to 5,2% in 2011. The proportion of households with access to piped water, increased from 62,3 % in 2001 to 73,4% in 2011. Households that use electricity for lighting increased from 70,2% in 2001 to 84,7% in 2011, while households that use electricity for cooking increased from 52,2% to 73,9% over the same period. Households that used electricity for heating increased from 49,9% in 2001 to 58,8% in 2011. The number of households with a television set increased by 20,7% since 2001.
>
> (South Africa Government Online, 2014)

Without such improvements, health problems are likely to be exacerbated, especially with increased urbanization across the developing world.

Food quality as well as quantity remain important issues today. But as a Royal Society of Chemistry (2014) report on Britain states, commonly used additives in the 19th century were often poisonous. To whiten bread, bakers sometimes added alum and chalk to the flour, while mashed potatoes, plaster of Paris (calcium sulphate), pipe clay and sawdust were used to increase the weight of loaves. Rye flour or dried powdered beans replaced wheat flour, and the sour taste of stale flour could be disguised with ammonium carbonate. Brewers often added mixtures of bitter substances, some containing poisons like strychnine, to 'improve' the taste of the beer and save on the cost of hops. By the beginning of the 19th century, the use of such substances in manufactured foods and drinks was so common that town dwellers had begun to develop a taste for adulterated foods and drinks; white bread and bitter beer were in great demand. Coffee was adulterated with acorns and burnt sugar and tea with used leaves, Prussian blue and copper salts. At the same time in the United States with the shift from an agricultural to an industrial economy and urbanization disconnecting people from food production, the debasement of food for profit became rampant. Milk was often watered down and coloured with chalk or plaster – substances which were also added to bulk up flour. Lead was added to wine and beer, and coffee, tea and spices were routinely mixed with dirt, sand or other leaves. Laws were enacted but were difficult to enforce until there were improvements in the routine testing of substances. Significant food quality legislation was passed in France (1851 and 1905), Britain (1860, 1872, 1875), Germany (1879), Belgium (1890), Austria (1896), Switzerland (1905), the United States (1906, 1938) and Spain (1908), but it was not until regulation and industrial food chemistry were linked and the courts understood what adulteration meant was the technological and legal progress used for health benefit. With the demand for warranties for foods to be safe, national and then international standards were developed for food safety, ultimately with the Codex Alimentarius Commission, set up in 1963 as a joint enterprise of the Food and Agriculture Organization and the World Health Organization. The main objective of the Codex is protecting the health of consumers and ensuring fair practices in the food trade by promoting coordinated international food standards (see Codex, 2016).

Adulteration and contamination of foods still occur often through poor hygienic practices but also through error and corruption. So in Spain in 1981 industrial oil was sold as cooking oil, resulting in 600 deaths. Through profiteering, milk and infant formula and other food materials and components were adulterated with melamine in Gansu, China, in 2008. An estimated 300,000 victims were reported, with six infants dying from kidney stones and other kidney damage and an estimated 54,000 babies being hospitalised. The chemical appeared to have been added to milk to cause it to appear to have a higher protein content. In a separate incident four years before, watered-down milk had resulted in 13 infant deaths from malnutrition. Furthermore, many low- and middle-income countries have low-quality food controls, while the food processing business is ensuring a taste for its products. In fact, technical advances in bottling, canning and heat treatment are at the basis of food processing. Other techniques such as drying, preserving, freezing and concentrating aided the development of consumer society while in themselves often being initially developed in military and space exploration contexts.

While there is concern with food additives and pesticide exposures, it is the changing content of diets – initiated by technological advances to use products and increase profit – which is of major concern. In an early report, Senauer (2003, p. 9) wrote,

> Americans have increased, in particular, their consumption of refined grains, in such products as pasta and tortilla chips, fats and oils, in such products as cheese and salad dressing, and caloric sweeteners, especially corn sweetener in such products as soft drinks. In each of these categories average consumption exceeds the recommendation for a healthy diet, while the typical American consumes too few whole grains, vegetables and fruits.

There is a significant debate about the role of fructose, corn syrup and added salt in diets in the industrialized world. Yet few agencies, including WHO and CDC, challenge the refined products of large food processing companies but emphasize a need for a balanced diet with sufficient physical activity. The solution is seen as lifestyle change not legislative action.

Benefits of fail-safe technologies

Technological fixes and design have been central to improvements with respect to death, injury or economic loss in the areas of transportation and energy production. It was some 30 years ago that Perrow (1984) argued that high-risk technologies might be fatally flawed and lead to normal accidents because of their being highly complex yet coupled systems. Organizational rather than operational failure might lead to what are called black swan events – rare and potentially catastrophic. Yet safety culture has tried to not only make risk reversible but non-existent in the first place. It is

> the product of individual and group values, attitudes, perceptions, competencies, and patterns of behaviour that determine the commitment to, and the style and proficiency of, an organization's health and safety management . . . Organizations with a positive safety culture are characterized by communications founded on mutual trust, by shared perceptions of the importance of safety and by confidence in the efficacy of preventive measures.
>
> (HSC, 1993)

Furthermore, Reason (1998, p. 294) suggests that an ideal safety culture is "the 'engine' that drives the system towards the goal of sustaining the maximum resistance towards its operational hazards". For environmental health research, perhaps the most interesting work has been carried out around nuclear safety, where black swans may be seen – rare events, especially ones that have never occurred – difficult to foresee, expensive to plan for and easy to discount with reductionist evidence. Yet these events occur, as the Chernobyl disaster of 1986 demonstrates. At Chernobyl, operator error, power surges, reactor vessel rupture and the exposing of graphite moderator to air led to a major radiation contamination, affecting Belarus, Ukraine, Russia and much of Europe. The disaster crippled the Soviet economy and led to contested mortality outcomes as well as serious illnesses, including different forms of cancer (see also Chapters 5 and 9).

Yet as nuclear designers comment, the event earlier at Three Mile Island led to design changes which minimized the impact of the Fukushima nuclear disaster. Yet at Fukushima, the risks faced had not been well analyzed. The operating company was poorly regulated and did not know what was going on. The operators made mistakes. The representatives of the safety inspectorate fled. Some of the equipment failed. "The establishment repeatedly played down the risks and suppressed information about the movement of the radioactive plume, so some people were evacuated from more lightly to more heavily contaminated places" ("Blow-Ups Happen," 2012). Yet design improvements and the establishment of fail-safe mechanisms may ensure that other communities are more protected. As Butler (2011) notes, the KANUPP plant in Karachi, Pakistan, has 8.2 million living within a 30-km radius, although it has just one relatively small reactor with an output of 125 megawatts. But larger plants – Taiwan's 1,933-megawatt Kuosheng plant with 5.5 million people within a 30-km radius and the 1,208-megawatt Chin Shan plant with 4.7 million; both zones include the capital city of Taipei – are also worrying. Overall, 21 nuclear plants have populations larger than 1 million within a 30-km radius, and six plants have populations larger than 3 million within that radius. On the other hand, several countries in the EU have put a freeze on new nuclear plant construction and still others are moving away from nuclear power altogether due to the potential for another Chernobyl- or Fukushima-style disaster ("When the Steam Clears," 2011). Design fixes can also alter the distribution of risk, say between car drivers and pedestrians, as we have shown. In fact, one of the major ways of reversing risk is through pollution control, although there is often a battle between economic development (and jobs) and environmental protection (and health). In the recent U.S. Congressional debate about clean air, the benefits of pollution control were noted. Improvements in air quality have direct positive effects on health. Improvements in air quality in preparation for the 1996 Atlanta Olympics led to significantly lower rates of childhood asthma events, including reduced emergency department visits and hospitalizations. Children in southern California who moved to communities with higher air pollution levels had lower lung function growth rates than children who moved to areas with lower air pollution levels (see Li et al., 2012). Regulation in fact saves lives, which also tends to result in net economic benefits (see Table 6.1).

Table 6.1 Impact of air pollution rules on premature deaths, hospitalizations, asthma aggravation cases, including net economic benefits

Air pollution rule	*Premature deaths avoided yearly*	*Hospitalisations avoided yearly*	*Asthma aggravation cases avoided yearly*	*Net economic benefits yearly (in billions $)*
Reduction of mercury, air toxins from power plants	11,000	5,700	130,000	$37 to $90
Reduction of mercury, air toxics from industrial boilers and incinerators	8,100	5,100	52,000	$27 to $80
Reduction of mercury, air toxics from cement kilns	2,500	1,740	17,000	$.57 to $17
Total	21,600	12,540	199,000	$70–$187

Note: Premature death figures are proposed maximum number of fatalities avoided each year.

Source: Adapted from EPA (2013).

Air quality standards, emission standards for vehicles and other technical actions to reduce air emissions were tightened during the 1990s in Mexico City, contributing to downward trends of carbon monoxide, nitric oxide and nitrogen dioxide (NOx) and ozone levels. Levels of emissions were reduced by half at some sites, resulting in an estimated reduction of 3,000 to 6,000 excess deaths (see Jamison et al., 2006). Tengs et al. (1995) estimated the relatively low cost of pollution control for lives saved, although some appeared to be extremely expensive.

Resilience – ways of coping

Technological changes are largely societal ways of preventing or reversing risk situations. On an individual level, much attention has been paid to the impact of environmental challenges to psycho-social health and how they might cope with any environmental anxieties. We have noted that individuals use a series of heuristics through which they evaluate situations to be risky or not. Furthermore, there has been attention paid to coping strategies. Much work has been based on Lazarus and Folkman's (1984) elaboration that the individual appraises the situation cognitively. There are two types of appraisals, primary and secondary. A primary appraisal is made when the individual makes a conscious evaluation of whether a situation leads to a harm or a loss, a threat or a challenge. Then secondary appraisal takes place when the individual asks him/herself "What can I do?" by evaluating the coping resources around him/her. These resources include physical resources, such as how healthy one is or how much energy one has, social resources, such as the family or friends one has to depend on for support in his/her immediate surroundings, psychological resources, such as self-esteem and self-efficacy, and also material resources such as how much money you have or what kind of equipment you might be able to use. We shall examine the likely effect of this material resource distribution in the next chapter. Suffice it to say that individuals can remove a stressful situation by moving away or changing their present environment. Evans and Cohen (1987) argue that stressors in the environment can be multiplicative, making coping challenging especially if control and the search for predictability are important dimensions of coping. In some instances, the stress is anticipatory – about an event that might happen, so fearing the worst, risk anxiety is heightened. It is possible that if the event occurs – as in the planning and location of a waste site – reappraisal occurs and some of the anticipatory fears are removed (see Elliott et al., 1997). If adaptation, however, moves beyond the individual it can lead to strategies of resilience to prevent or reverse risk.

Through these adaptations and ways of coping, individuals and communities may not only show vulnerability to environmental problems but also resilience in such situations. Resilience is seen as a "measure of the persistence of systems and their ability to absorb change and disturbance and still maintain the same relationships between populations or state variables" (Holling, 1973, p. 14). Cutter et al. (2008), in the context of natural disaster, construct a resilience model as depicted in Figure 6.3.

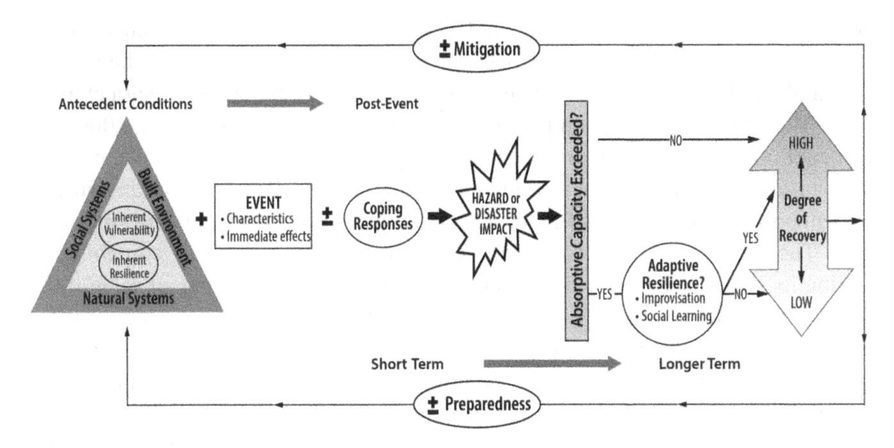

Figure 6.3 Model of resilience in the context of a natural disaster

Source: Cutter et al. (2008), with kind permission of Elsevier.

Resilience is determined by organizational and infrastructural factors but also largely by social ones in the sense of how communities respond to potential risk states or reverse or mitigate against risk events.

Resilience requires planning and/or the availability of resources. These resources can be within communities, often in the form of social capital, as in the existence of social ties and networks, feelings of trust and norms of reciprocity and availability of a range of resources – libraries, recreation centres, service organizations. Sampson et al. (2002) argue that neighbourhood residents can use these in adapting to or being resilient about social and environmental crises. If such community capacity is available and influential it can lead to strengthened resilience in times of crisis. Much work has been geared through using this capacity to improve the lives of children. Chaskin (2008) documents how low-income multi-ethnic neighbourhoods in Chicago regrouped police and service agencies to provide non–gang-related activities, while Latinos in the same city used their capacity to address vacant and derelict land as well as improve local infrastructure.

Resilience can be worn out and may be subject to surprises. Gunderson (2003) refers to different types of surprise: local surprise, unexpected discrete events at a small scale; cross-scale surprise, discontinuities in long-term trends over regional and global scales; and novel surprises, disturbances that are unique or not previously experienced by the social-ecological system and produce unpredictable consequences. Surprise is of course in the nature of things. It is about depth as well as surface (see Giddens, 1991). It may bring novelty to adaptation and resilience as individuals and communities attempt to find control, security and normality. Yet surprise is as inevitable as the uncertainty it helps produce. Industrial accidents often come as a surprise because of the technological systems developed to minimize risk. There should, therefore, be no expectation of a gas leak, a fire or an explosion. But they happen and are surprises. Perhaps the great one was the Bhopal disaster when in 1984, more than 40 tons of methyl isocyanate (MIC) gas leaked from a pesticide plant, immediately killing between 7,000 and 10,000 in the first three years and up to 20,000 by 2003 (Mishra et al., 2009). More than half a million people were exposed to MIC; around 120,000 of them continue to suffer from chronic respiratory, ophthalmic, reproductive, endocrine, gastrointestinal, musculo-skeletal, neurological and mental disorders. Post-accident pollution continues as materials left behind seep into the soil and groundwater. The company, Union Carbide (UC), refused to recognize its culpability, blaming its Indian subsidiary or sabotage by workers while fighting all scientific data (see Broughton, 2005). Legal and clean-up wrangles continue, as does the toxic load on residents (see Ansell and Tinsley, 2011). ICJB (2014) continues to report on Dow Chemicals – the now owner of UC – reluctance to act as it prepares for the 30th anniversary. This surprise of an apparently fail-safe system should bring greater compliance, better design and greater oversight to the production of toxic products. It has not, as the fires and explosions in oil terminals, pesticide plants and fertilizer factories in the UK and U.S. attest. Failure of fail-safes leads then to surprise and usually inequitable suffering among workers and usually the surrounding blue-collar communities.

References

Andreoni, J. 1995. Warm-glow versus cold-prickle: The effects of positive and negative framing on cooperation in experiments. *The Quarterly Journal of Economics*, 110(1), 1–21.

Ansell, R., & Tinsley, A. 2011. *Bhopal's Never Ending Disaster*. Available at: http://www.environmentalistonline.com/article/2011–10–13/bhopal-s-never-ending-disaster (Accessed 15 April 2015).

Bloom, D.E., Canning, D., & Sevilla, J. 2004. The effect of health on economic growth: A production function approach. *World Development*, 32(1), 1–13.

Blow-Ups Happen. 2012. *The Economist*. Available at: http://www.economist.com/node/21549095 (Accessed 11 March 2015).

Broughton, D. 2005. The Bhopal disaster and its aftermath: A review. *Environmental Health*, 4, 6.

Butler, D. 2011. Reactors, residents and risk. *Nature*. Available at: http://www.nature.com/news/2011/110421/full/472400a.html (Accessed 11 March 2015).

Chaskin, R. 2008. Resilience, community, and resilient communities: Conditioning contexts and collective action. *Child Care in Practice*, 14, 65–74.

Codex alimentarius. 2016. FAO, Geneva. http://www.fao.org/fao-who-codexalimentarius/about-codex/en/

Concerned Citizens for Nuclear Safety. 2003. *New Mexico's Right to Know: The Impacts of Los Alamos National Laboratory Operations On Public Health and the Environment*. Available at: http://www.nuclearactive.org/docs/RighttoKnow.pdf (Accessed 4 March 2015).

Cutter, S.L., Barnes, L., Berry, M., Burton, C., Evans, E., Tate, E., & Webb, J. 2008. A place-based model for understanding community resilience to natural disasters. *Global Environmental Change*, 18(4), 598–606.

Dales, R.E., Zwanenburg, H., Burnett, R., & Franklin, C.A. 1991. Respiratory health effects of home dampness and molds among Canadian children. *American Journal of Epidemiology*, 134(2), 196–203.

Dolk, H., Vrijheid, M., Armstrong, B., Abramsky, L., Bianchi, F., Garne, E., . . . & Tenconi, R. 1998. Risk of congenital anomalies near hazardous-waste landfill sites in Europe: The EUROHAZCON study. *The Lancet*, 352(9126), 423–427.

Edge, S., & Eyles, J. 2013. Message in a bottle: Claims disputes and the reconciliation of precaution and weight-of-evidence in the regulation of risks from Bisphenol A in Canada. Health, Risk & Society, 15(5), 432–448.

Elliott, S.J., Taylor, S.M., Hampson, C., Dunn, J., Eyles, J., Walter, S., & Streiner, D. 1997. 'It's not because you like it any better . . .': Residents' reappraisal of a landfill site. *Journal of Environmental Psychology*, 17, 229–241.

EPA. 2010. *Nonpoint Source Pollution (Polluted Runoff)*. Available at: http://www.epa.gov/region2/water/npspage.htm (Accessed 30 March 2015).

EPA. 2013. *Cross-State Air Pollution Rule (CSAPR)*. Available at: http://www.epa.gov/airtransport/CSAPR/ (Accessed 20 March 2015).

EPA. 2014. *Draft Report on the Environment*. Available at: http://www.epa.gov/ncea/roe/index.htm (Accessed 2 March 2015).

Estry, D.C., & Moffa, A.L. 2012. Why climate change collective action has failed and what needs to be done within and without the trade regime. *Journal of International Economic Law*, 15(3), 777–791.

Evans, G., & Cohen, S. 1987. 'Environmental Stress' in Stokols, D. & Altman, I. (eds.). *Handbook of Environmental Psychology*. New York: Wiley. pp 571–610.

Fischer, M., & Lotz, S. 2014. Is soft paternalism ethically legitimate? – The relevance of psychological processes for the assessment of nudge-based policies. *CGS Working Paper*, 5, no. 2. Cologne: Cologne Graduate School in Management, Economics and Social Sciences.

Giddens, A. 1991. *Modernity and Self-Identity*. Stanford: Stanford University Press.

Gillis, J. 2012. *In Poll, Many Link Weather Extremes to Climate Change*. Available at: http://www.nytimes.com/2012/04/18/science/earth/americans-link-global-warming-to-extreme-weather-poll-says.html?_r=0 (Accessed 2 April 2015).

Gunderson, L. 2003. 'Adaptive Dancing: Social Resilience and Ecological Crises.' in Berkes, F., Colding, J., & Folke, C. (eds.). *Navigating Social-Ecological Systems*. Cambridge: Cambridge University Press. pp 33–52.

Haines, A., Bruce, N., Cairncross, S., et al. 2012. Promoting health and advancing development through improved housing in low-income settings. *Journal of Urban Health*, 90, 810–831.

Hamilton, C. 2007. *Scorcher: The Dirty Politics of Climate Change*. Collingwood: Black Inc. Agenda.

Health Canada. 2014. *Bisphenol A*. Available at: http://www.hc-sc.gc.ca/fn-an/securit/packag-emball/bpa/index-eng.php (Accessed 10 April 2015).

Health and Safety Commission (HSC). 1993. *Third Report: Organizing for Safety*. ACSNI Study Group on Human Factors. London: HMSO.

Holling, C.S. 1973. Resilience and stability of ecological systems. *Annual Review of Ecology and Systematics*, 4, 1–23.

International Campaign for Justice in Bhopal. 2014. *Dow's Reluctant Legacy*. Available at: http://www.bhopal.net/what-happened/3-the-aftermath-1985-present/dow-chemicals-reluctant-legacy (Accessed 10 April 2015).

Jamison, D.T., Breman, J.G., Measham, A.R., Alleyne, G., Claeson, M., Evans, D.B., . . . & Musgrove, P. 2006. *Disease Control Priorities in Developing Countries*. Washington, DC: World Bank.

Kahneman, D., & Knetsch, J.L. 1992. Valuing public goods: The purchase of moral satisfaction. *Journal of Environmental Economics and Management*, 22(1), 57–70.

Krieger, J., & Higgins, D.L. 2002. Housing and health: Time again for public health action. *American Journal of Public Health*, 92(5), 758–768.

Kulldorff, M., Athas, W.F., Feurer, E.J., Miller, B.A., & Key, C.R. 1998. Evaluating cluster alarms: A space-time scan statistic and brain cancer in Los Alamos, New Mexico. *American Journal of Public Health*, 88(9), 1377–1380.

Lazarus, R.S., & Folkman, S. 1984. *Stress, Appraisal, and Coping*. New York: Springer.

Local Government Association. 2013. *Changing Behaviours in Public Health: To Nudge or to Shove?* London: LGA.

Li, J., Ewart, G., Kraft, M., & Finn, P.W. 2012. The public health benefits of air pollution control. *Journal of Allergy and Clinical Immunology*, 130(1), 22–23.

Luginaah, I., Smith, K., & Lockridge, A. 2010. Surrounded by Chemical Valley and 'living in a bubble': The case of the Aamjiwnaang First Nation, Ontario. *Journal of Environmental Planning and Management*, 53(3), 353–370.

Mackenzie, C.A., Lockridge, A., & Keith, M. 2005. Declining sex ratio in a first nation community. *Environmental Health Perspectives*, 113(10), 1295–1298.

Michaels, D. 2008. *Doubt Is Their Product*. New York: Oxford University Press.

Mishra, P., Samarth, R., Pathka, N., et al. 2009. Bhopal Gas Tragedy: Review of clinical and experimental findings after 25 years. *International J Occupational Medicine and Environmental Health*, 22, pp 193–202.

Nudge Nudge, Think Think. 2012. *The Economist*. Available at: http://www.economist.com/node/21551032 (Accessed 3 March 2015).

Perrow, C. 1984. *Normal Accidents*. New York: Basic Books.

Popkin, B.M., Kim, S., Rusev, E.R., Du, S., & Zizza, C. 2006. Measuring the full economic costs of diet, physical activity and obesity-related chronic diseases. *Obesity Reviews*, 7(3), 271–293.

Price, V. 2011. *The Orphaned Land*. Albuquerque: University of New Mexico Press.

Quigley, M. 2013. Nudging for health: On public policy and designing choice architecture. *Medical Law Review*, 21, 588–621.

Ravishankara, A.R., Daniel, J.S., & Portmann, R.W. 2009. Nitrous oxide (N2O): The dominant ozone-depleting substance emitted in the 21st century. *Science*, 326(5949), 123–125.

Reason, J. 1998. Achieving a safe culture: Theory and practice. *Work and Stress*, 12, 293–306.

Riffkin, R. 2014. *Climate Change Not a Top Worry in U.S. Gallup Politics*. Available at: http://www.gallup.com/poll/167843/climate-change-not-top-worry.aspx (Accessed 2 April 2015).

Rosen, G. 1958. *A History of Public Health*. New York: MD Publications.

Rowland, F. 2007. *Stratospheric Ozone Depletion by Chlorofluorocarbons (Nobel Lecture)*. Available at: http://www.eoearth.org/view/article/156270/ (Accessed 4 April 2015).

Royal Society of Chemistry. 2014. *The Fight against Food Adulteration*. Available at: http://www.rsc.org/education/eic/issues/2005mar/thefightagainstfoodadulteration.asp (Accessed 20 March 2015).

Rydin, Y., & Pennington, M. 2000. Public participation and local environmental planning: The collective action problem and the potential of social capital. *Local Environment*, 5(2), 153–169.

Sampson, R.J., Morenoff, J.D., & Gannon-Rowley, T. 2002. Assessing 'neighborhood effects': Social processes and new directions in research. *Annual Review of Sociology*, 28, 443–478.

Senauer, B. 2003. The Obesity Crisis: Challenge to the Food Industry. *Working Paper 03–04*. The Food Industry Center University of Minnesota. Available at: http://ageconsearch.umn.edu/bitstream/14309/1/tr03–04.pdf (Accessed 2 April 2015).

Siegal, G., Siegal, N., & Bonnie, R.J. 2009. An account of collective actions in public health. *American Journal of Public Health*, 99(9), 1583–1587.

South Africa government online. 2014. *Housing*. Available at: http://www.gov.za/aboutsa/housing.htm (Accessed 2 April 2015).

Stern, P.C. 2000. Psychology and the science of human-environment interactions. *American Psychologist*, 55(5), 523–530.

Suarez, L., Felkner, M., Brender, J.D., Canfield, M., Zhu, H., & Hendricks, K.A. 2012. Neural tube defects on the Texas-Mexico border: What we've learned in the 20 years since the Brownsville cluster. *Birth Defects Research Part A: Clinical and Molecular Teratology*. 94(11), 882–892. doi: 10.1002/bdra.23070.

Sweet, S.N., & Fortier, M.S. 2010. Improving physical activity and dietary behaviours with single or multiple health behaviour interventions? A synthesis of meta-analyses and reviews. *International Journal of Environmental Research and Public Health*, 7(4), 1720–1743.

Tengs, T.O., Adams, M.E., Pliskin, J.S., Safran, D.G., Siegel, J.E., Weinstein, M.C., & Graham, J.D. 1995. Five-hundred life-saving interventions and their cost-effectiveness. *Risk Analysis*, 15(3), 369–390.

Van Larebeke, N.A., Sasco, A.J., Brophy, J.T., Keith, M.M., Gilbertson, M., & Watterson, A. 2008. Sex ratio changes as sentinel health events of endocrine disruption. *International Journal of Occupational and Environmental Health*, 14(2), 138–143.

Walsh, B. 2010. *The Perils of Plastic*. Available at: http://content.time.com/time/specials/packages/article/0,28804,1976909_1976908_1976938–1,00.html (Accessed 1 March 2015).

When the Steam Clears. 2011. *The Economist*. Available at: http://www.economist.com/node/18441163 (Accessed 20 February 2015).

WHO. 2006. *Preventing Disease through Healthy Environments*. Geneva: Author.

WHO. 2014a. *Economic Growth*. Available at: http://www.who.int/trade/glossary/story019/en/ (Accessed 2 March 2015).

WHO. 2014b. *Obesity and Overweight, Fact Sheet 311*. Available at: http://www.who.int/mediacentre/factsheets/fs311/en/ (Accessed 2 April 2015).

World Economic Forum. 2014. *Global Risks 2014*. Geneva: Author.

7 Delayed risk, exposures and health outcomes

Introduction

Concern over environmental risks is often significant if the impact of the event or phenomenon is not immediate. Impact can often be delayed. For some substances, delay heightens concern as in the case of nuclear waste and various cancers, but for others such delay may lead people to ignore the potential health impacts as in the case of solar radiation and skin cancer. For example, much attention has been paid to the chemical constituents of household cleaning and health protection products. Triclosan is used as an antibacterial and must be labelled if used in products. There is some evidence that it is toxic and a suspected endocrine disrupter that can mimic or interfere with the function of hormones (see Lee et al., 2014). The European Union classifies triclosan as irritating to the skin and eyes and as very toxic to aquatic organisms, noting that it may cause long-term adverse effects in the aquatic environment. It is also regulated by the FDA due in part because triclosan can also react in the environment to form dioxins. There has been pressure from civil society which has led Minnesota to ban its use. Other jurisdictions still regard current exposure levels as safe, but recent toxicological work involving mice has shown that triclosan can disrupt the expression of genes associated with breast cancer progression using in vitro and in vivo models. Furthermore treatment with the chemical increased MCF-7 breast cancer cell proliferation more than twofold in vitro compared to controls. Such findings, if replicated, create uncertainty over the impact of long-term exposure to triclosan. And such delays can be troubling especially if it appears economics or politics trump health concerns.

The case of asbestos provides some insights into delay, as political complexities reduced regulatory control. In 2011 Canada opposed listing chrysotile asbestos as a hazardous chemical for the Rotterdam Treaty and remained silent at home by continuing to ship asbestos to developing nations in particular (CBC, 2013). Also opposing the inclusion were Vietnam, Kazakhstan and Kyrgyzstan, with no inclusion meaning recipients need not be informed of possible health effects. Canada's position – brought about by its support for the national asbestos industry – remains that chrysotile asbestos can be used safely 'under controlled conditions'. And while its use is banned under national law, Canada in 2015 still refuses to support listing chrysotile asbestos as a hazardous substance under global trade rules. In so doing, the Canadian government will help Russia continue exporting asbestos to developing countries with no safety controls required (Canada Is On the Sidelines, 2015). This apparent double standard comes despite all the evidence that long-term exposure results in severe health impacts. Over time, asbestos fibres accumulate in the lungs and cause scarring and inflammation. This makes breathing increasingly difficult and can even lead to cancer and other illnesses. Asbestos is listed by the International Agency for Research on Cancer (IARC) as a group 1 (cancerous to humans) carcinogen – the highest. Symptoms of these diseases may not appear until 10 to 50 years after the initial exposure occurred. These delayed impacts affect mainly workers but there are also residential impacts. A 2009 study examined the effects of environmental exposure in a population living near an asbestos manufacturing plant, namely elevated rates of malignant pleural mesothelioma (MPM) and other asbestos-related conditions in Shubra El-Kheima, Egypt, an industrial city containing the Sigwart Company asbestos cement plant. It compared disease rates in individuals working in the plant, those living near the plant and those in a control group with no known asbestos exposure. In total, the study had more than 4,000 participants. The rate of MPM was highest in the group with environmental asbestos exposure, with 2.8% of this group reporting the cancer. The group with occupational exposure had a strikingly lower rate of only 0.8%. As expected, the control group had the fewest incidences, with a rate of 0.1%. These rates varied for other illnesses such as diffuse pleural thickening. Overall, the study found a slightly higher – but still comparable – rate of asbestos-related illnesses in asbestos workers compared to nearby residents. It should be noted the cement plant closed in 2004. Yet in 2013 a study reported that chrysotile asbestos remained relatively unproblematic if safeguards were in place. Others have noted that the primary author is often funded by the Quebec Chrysotile Institute, although it closed in 2011 and the Canadian province stopped support for the last mine in 2012 (see Madkour et al., 2009; Faith, 2013).

The desire for scientific certainty can also affect the likelihood of identifying delayed and long-term impacts. Thus the impact of methylmercury poisoning was significantly delayed. The failure to recognize the distinctive clinical features of serious methylmercury poisoning in adults delayed the identification of the aetiology of Minamata disease and thus recognition of the full extent of the Japanese outbreak. Even when methylmercury had been established as the chemical cause of the disease, strict diagnostic requirements and case definitions assumed that the disease was a characteristic all-or-nothing phenomenon, thereby excluding less distinctive cases and obscuring the dose–effect relationship.

In Chapter 1, we describe the toxicity of methyl mercury with specific reference to laboratory studies, but it is through epidemiologic studies that delayed effects in humans are detected and control policy implemented. Thus, delay also refers to political action on suspected substances amid scientific uncertainty. Minamata, Japan, was a poor fishing village which also housed from the 1930s a plastics manufacturing plant. Anecdotes of cats dying and people slurring their speech and dropping chopsticks appeared. These provided the first signs of what became a poisoned village as plant effluent polluted the water, a main source of protein through fish and shellfish. Severe physiological effects began to emerge, including successive loss of motor control, which was devastating and sometimes resulted in partly paralyzed and contorted bodies. Epidemiological evidence of mercury poisoning emerged in 1956, and the responsibility of the plant was determined in 1959. The company, Chisso, employer of 60% of the town workforce, denied this and provided false evidence. It also split into several companies to avoid blame. It was not found legally liable until 1972, ceasing to use mercury in 1968 (see Ishimure, 1990). The first likely cases of developmental methylmercury poisoning were described in 1952 (Engleson and Herner, 1952), replicated in laboratory animals in 1972 (Spyker et al., 1972), and the first prospective population study of prenatal methylmercury toxicity due to contaminated seafood in humans was published in 1986 (Kjellström et al., 1986). Such findings were subsequently summarized in a report from a study group on Minamata disease (Social Scientific Study Group on Minamata Disease, 1999; Harada, 2004). The Japanese government (2002) certified 2,265 victims, nearly 1,800 of whom had died. However, scientific consensus on prenatal vulnerability was hampered by focusing on uncertainties in the evidence, and international agreement on the need for protection against prenatal exposures was reached only in 2003. After the publication of new data on the adverse effects of low-level exposures to methylmercury, regulatory agencies requested further scientific scrutiny. Expert committees emphasized uncertainties and weaknesses in the available data. Less attention was paid to the question of what could have been known, given the research methods and possibilities, and whether developmental neurotoxicity at low methylmercury doses could be ruled out. The reports also generally ignored that measurement imprecision most likely resulted in an underestimation of the true effects. Instead, more research was recommended. The insistence on solid evidence promoted by polluters and regulatory agencies therefore agreed with a desire among researchers to expand scientific activities in this area. However, the wish to obtain more complete proof had the untoward effect of delaying corrective action (see Grandjean et al., 2010). But in the case of Minamata, Chisso paid financial compensation to more than 10,000 people based on emotional as well as scientific arguments. Chisso, rather like any imperial Japanese entity, agreed to repay all victims fully at a cost of more than $50 million. The company has been aided by local and provincial governments, which have waived over half a billion dollars in liabilities (see "New Minamata Compensation," 2010).

So what is delay? The role of latency

Latency reflects the time lag between an exposure and the result outcome. It can vary in time from seconds to years depending on the nature and type of exposure. Some environmental exposures for cancer, for example, have a long latency period. The latency for pleural mesothelioma is 20 to 50 years, a disease that typically develops within the chest cavity and lungs. The most common type of malignant mesothelioma, it often spreads to numerous other organs and lymph nodes in the body. The difficulty in its diagnosis and detection makes treatment tougher. Symptoms such as chest pains, weight loss and fever are far too common to be noticed as a precursor for cancer. A study of diagnosed patients found the latency to be 48 years for women and 53 for men (see Haber and Haber, 2011). Mesothelioma is only one adverse outcome from asbestos exposure: there are also lung cancers, respiratory tract problems and gastrointestinal illnesses. Asbestos has been used widely by manufacturers and builders because of its desirable physical properties: sound absorption, average tensile strength, its resistance to fire, heat, electrical and chemical damage and affordability. It was used in such applications as electrical insulation for hotplate wiring and in building insulation. When asbestos is used for its resistance to fire or heat, the fibres are often mixed with cement or woven into fabric. It became widely used and advertised as a magical material for household use (see Gee and Greenberg, 2002). Health effects were noted in industrial settings as early as the 1920s. Bartrip (1998) notes that there was not an industrial coverage but a result of large-scale commercial exploitation of asbestos being still relatively new, and the long latency of the disease led it to be confused with tuberculosis. Furthermore the long latency led to the perception that workers were healthy until old age, the lack of evidence on cancer incidence led to a type 2 error (falsely concluding there was no effect – i.e., a false negative), which was exacerbated by the coincidence of carcinogenicity of asbestos and smoking in the 1950s (see Gee and Greenberg, 2002). As we have seen, the industrial and residential utility continued to be argued by particular states as sufficient benefit to counteract increasingly certain risk into the 21st century. But as a recent RAND (2014) report noted, since there is no easy way to establish first exposure, time lags are hard to determine with much precision. With banning asbestos, a slow process of bereavement, whistle-blowing and scientific evidence began to challenge even the utility of the product and slow court procedures (see Roselli, 2014).

Long time lags and the uncertain relationship between exposure and cancer outcome make attribution to environmental exposures difficult in non–smoker-based cancer epidemiology. In a 2005 review, Clapp et al. noted a wide range of associations between for example metals such as arsenic and cancers of the bladder, lung and skin; chlorination by-products such as trihalomethanes and bladder cancer; natural fibres such as asbestos and cancers of the larynx, lung, mesothelioma and stomach; petrochemicals and combustion products, including motor vehicle exhaust and polycyclic aromatic hydrocarbons and cancers of the bladder, lung and skin; pesticide exposures and cancers of the brain, Wilms tumour, leukemia and non-Hodgkin's lymphoma; reactive chemicals such as vinyl chloride and liver cancer and soft tissue sarcoma; and solvents such as benzene and leukemia and non-Hodgkin's lymphoma; tetrachloroethylene and bladder cancer; and trichloroethylene and Hodgkin's disease, leukemia and kidney and liver cancers. For example, they conclude that the evidence linking metal exposure from arsenic with bladder cancer is strong and extensive. Much of the evidence comes from epidemiologic studies conducted in regions with high concentrations of inorganic arsenic contaminants in drinking water. Several volatile chemicals have been linked with bladder cancer. Evidence from multiple studies examining chlorination by-products has consistently found elevated risk of bladder cancer, especially among populations with long-term exposure to chlorinated water (see also Letašiová et al., 2012). Bladder cancer develops slowly over time. It is largely a disease of older people, with the American Cancer Society reporting that the average age of detection is 73. Thus we can argue that besides any genetic predisposition, identifying and measuring known risk factors for individuals are massive, uncertain tasks. For example, where did drinking water come from, how long was the person resident in that area, how was it treated and what other modifying factors might there be? Furthermore, risks versus benefits come into play with respect to drinking water treatment since chlorination is an inexpensive and effective way of neutralizing micro-organisms which cause potentially fatal gastrointestinal illnesses, for example *E. coli*, which caused the Walkerton, Ontario, disaster. Without effective chlorination, water may become so polluted that people become seriously ill or die. There is then the need of a risk trade-off between elevated bladder cancer risk and possibilities of widespread gastric infections.

The focus on latency periods tends to be on very long latencies; they are hard to determine and quantify, particularly when the exposure is diffuse and even when the latency ends up being rather short. Yet a great deal of effort has been undertaken to assess the latency period between air pollution and respiratory illnesses and death.

Bell and Davis (2001) re-analyze the circumstances around the London Great Smog of 1952–1953. Public health insurance claims, hospital admission rates for cardiac and respiratory disease, pneumonia cases, mortality records, influenza reports, temperature and air pollutant concentrations are analyzed for December through February 1952 through 1953 and compared with those for the previous year or years. Pollution levels during the London smog were 5 to 19 times above current regulatory standards and guidelines (see Figure 7.1). Mortality rates for the smog episode were 50 to 300% higher than the previous year (see Figure 7.2). Claims that the smog only elevated health risks during and immediately following the peak fog of 5 to 9 December 1952 and that an influenza epidemic accounted fully for persisting mortality increases in the first two months of 1953 were rejected. They estimated about 12,000 excess deaths occurred. For later impacts, Thurston et al. (1989) estimated the time lag between pollution event and mortality was two days. This level of pollution caused by cold, inversion, the use of coal and the dumping of soot spurned a parliamentary committee to act quickly, and a Clean Air Act was passed in 1956. These fogs acted as harvesters of the sick and those with pre-existing conditions. For heart disease, two large and global pooled analyses have shown epidemiologically the link between ambient air pollution and incident myocardial infarction, heart failure admissions and mortality. Furthermore, the strongest effect of gaseous and particulate pollution resulting in heart failure admission or mortality was seen immediately at lag 0 days, with marked attenuation at longer lag period (see Shah and Newby, 2014). Air pollution remains a significant public health issue. Burnett et al. (1997) have noted the association with summer ozone and respiratory hospitalizations with a very

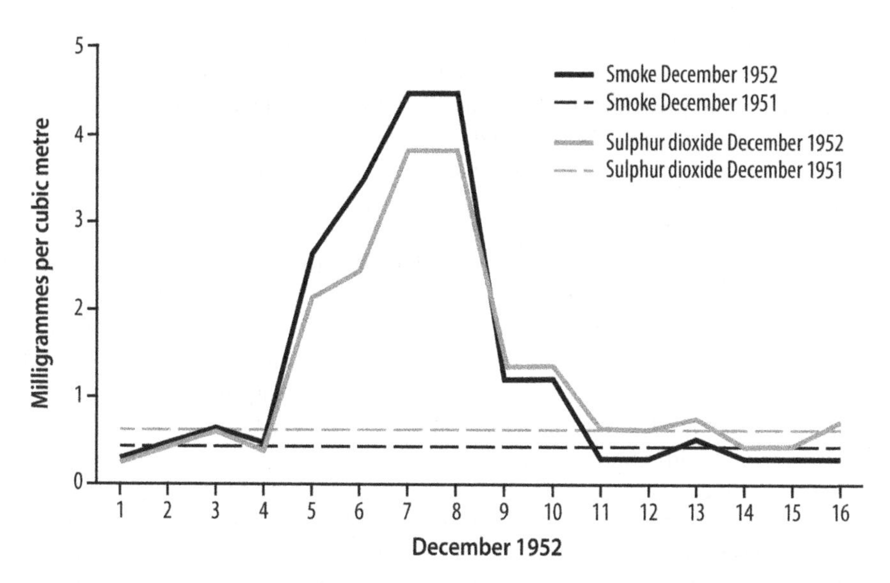

Figure 7.1 Smoke and sulphur dioxide concentrations at County Hall during the London smog of December 1952 compared to the average concentrations in December 1951

Source: Mayor of London (2002), under OGL.

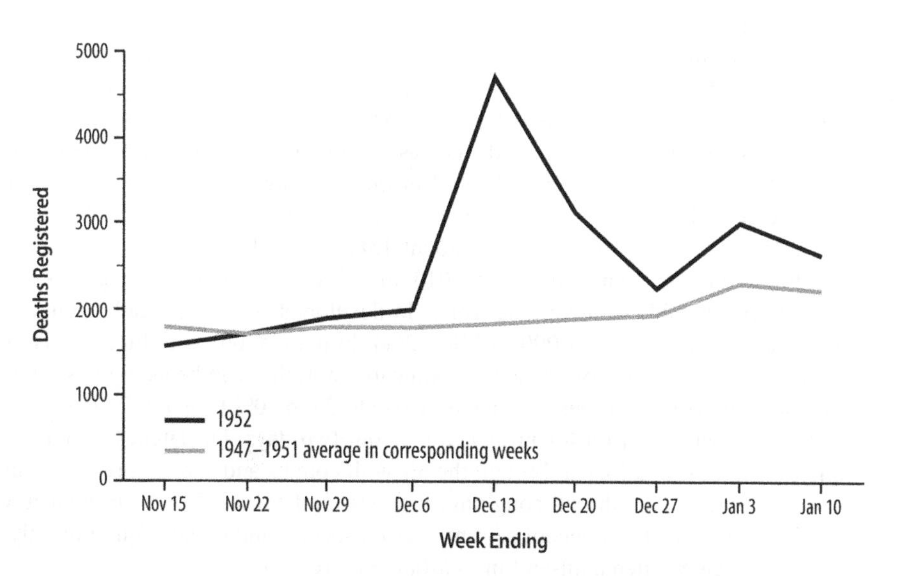

Figure 7.2 Deaths registered during the London smog of December 1952

Source: Mayor of London (2002), under OGL.

short time lag. Braga et al. (2001) noted that a 10-microgram/m^3 increase in the two-day mean of PM_{10} was associated with a 0.7% (95% CI: 0.3, 1.1) increase in deaths from myocardial infarction. When the distributed lag was assessed, two different patterns could be observed: respiratory deaths were more affected by air pollution levels on the previous days, whereas cardiovascular deaths were more affected by same-day pollution. Small increases in pollution matter given the very short latency period.

Latency periods are also important with respect to infectious diseases, in which they are often called the incubation period, which is the time elapsed between exposure to a pathogenic organism when symptoms and signs are first apparent. In a typical infectious disease, incubation period signifies the period taken by the multiplying organism to reach a threshold necessary to produce symptoms in the host. Measles, for example, takes 9 to 12 days to incubate, and diagnosis is initially difficult because of early non-specific symptoms. Once the lag is known (along with such things as the infectious period), identifying likely exposure sites and other potentially infected people becomes much easier for controlling disease spread. Thus, knowing that the Ebola virus has an incubation period from 2 to 21 days meant several people exposed to infectious patients had to be quarantined up to three weeks. The disease spreads easily through contaminated droplets in the air such that its diffusion takes a contagious form. As infection provides immunity, the virus needs new hosts to survive. Measles is therefore most sustainable in crowded districts with high birth rates, low-quality infrastructure, low nutrition and under-resourced health care systems. Vaccines are available to prevent its spread, but herd immunity is required, meaning that 95% of a community must be vaccinated. In several Western countries, discredited science, insistence of freedom of choice and the free-rider problem have led to reductions in take-up of many vaccines, including those for measles, mumps, chickenpox and whooping cough. New cases have begun to emerge. This is a concern as since 2001, U.S.–reported measles incidence has remained below 1 case per 1,000,000 population (see Papania et al., 2014). But measles largely remains an indicator of poverty and low levels of housing and health service development. In 2011, the WHO estimated that there were about 158,000 deaths caused by measles, a fourfold reduction from 1990. Yet in 2013, measles remained the leading cause of vaccine-preventable deaths in the world. In developed countries, death occurs in 1 to 2 cases out of every 1,000 (0.1%–0.2%). In populations with high levels of malnutrition and a lack of adequate healthcare, mortality can be as high as 10%. In cases with complications, the rate may rise to 20 to 30% (see WHO, 2014).

The incubation period for influenza is about two days, but latency is not the only concern. It is diffused through the air with coughs and sneezes, resulting in fever, headache, sore throat, coughing, tiredness and nausea. Yet the influenza A virus is of the greatest concern as it may cross species and mutates quite rapidly, making vaccines often a hit-and-miss affair (Figure 7.3).

Of great environmental health interest are influenza pandemics, which occur when a new strain of the influenza virus is transmitted to humans from another animal species. Species important in the emergence of new human strains are

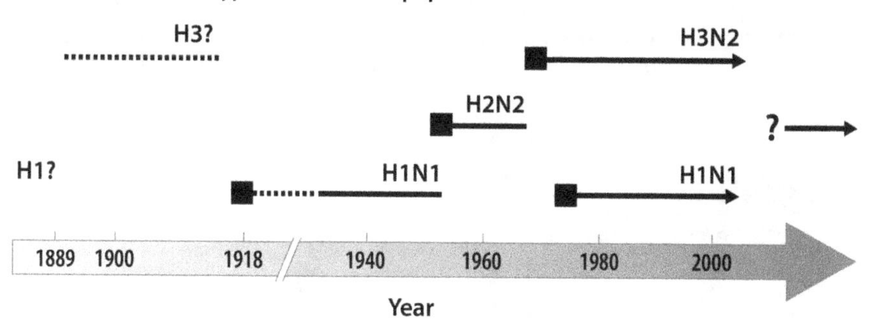

Figure 7.3 Variations in the influenza A virus in the human population
Source: Vasconcellos (2009).

pigs, chickens and ducks. These novel strains are unaffected by any immunity people may have to older strains of human influenza and can therefore spread extremely rapidly and infect very large numbers of people. There is therefore the ability of the virus to spread rapidly through a population weakened by close proximity to infected species and by poor nutrition, exhaustion and overcrowding (see Figure 7.4). Why the concern? Largely because of the Spanish flu, caused by an unusually severe and deadly influenza A virus strain of subtype H1N1. The Spanish flu pandemic lasted from 1918 to 1920, which started near the end of the WWI. Older estimates place the death toll at 40 to 50 million people, while current estimates suggest 50 million to 100 million people worldwide were killed, about 1% of global population (see Barry, 2005; Figure 7.5). To put this number in perspective, about 16 million people died from the war itself, while about 60 million died in WWII. This pandemic has been described as "the greatest medical holocaust in history" (see Potter, 2001). The death toll was proximally caused by an extremely high infection rate of up to 50% and the extreme severity of the symptoms, suspected to be caused by cytokine storms (see Patterson and Pyle, 1991). Haemorrhages were commonplace, and these along with the second infection of pneumonia killed many, unusually young adults, whereas the always vulnerable young and elderly were, and still are, the most common victims. Why this was the case for the Spanish flu remains an area of scientific debate, some pointing to possible lack of immunity from previous influenza exposures in this group. Others point to environmental and political conditions – what we might call distal, societal or upstream causes of the epidemic. These advocates argue that early outbreaks of a new disease with rapid onset and spreadability, high mortality in young soldiers in the British base camp at Etaples in Northern France in the winter of 1917 may be, at least to date, the most likely focus of origin of the pandemic. Pathologists working at Etaples and Aldershot barracks later agreed that these early outbreaks in army camps were the same disease as the infection wave of influenza in 1918. The Etaples camp had the necessary mixture of factors for

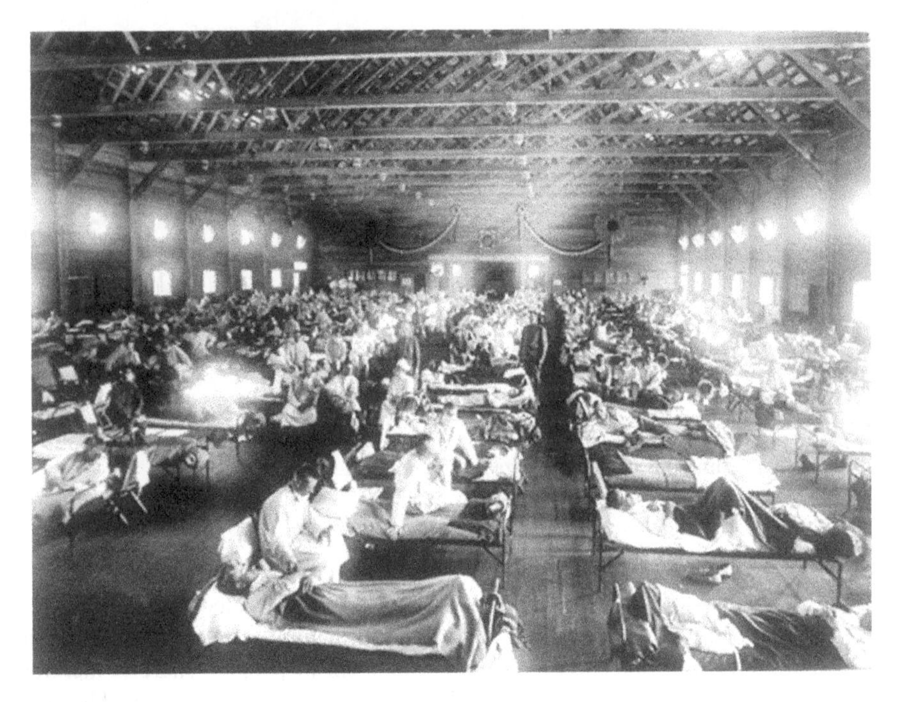

Figure 7.4 Influenza ward, Fort Riley, Illinois, 1917

Source: U.S. Government (1917).

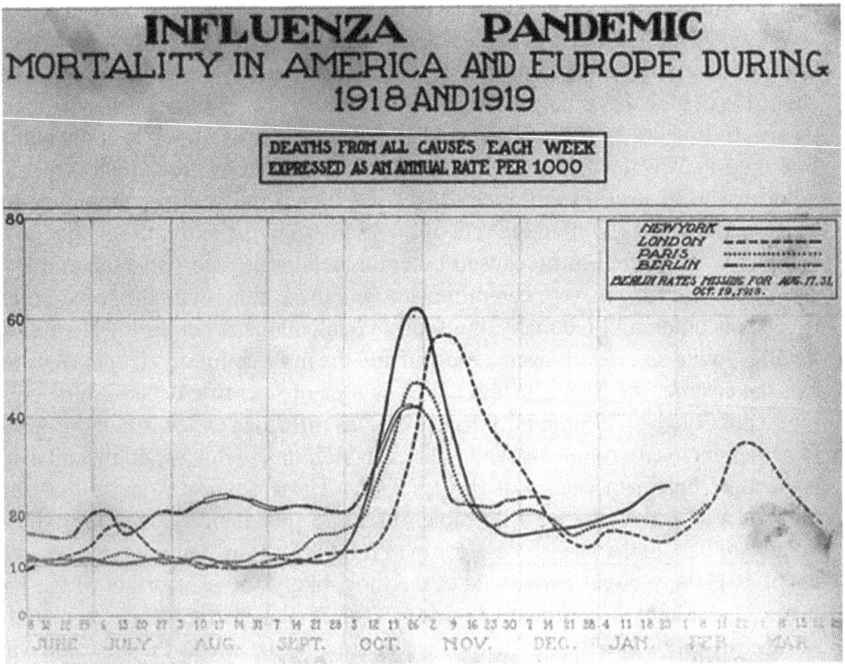

Figure 7.5 Influenza mortality between June 1918 and March 1919

Source: National Museum of Health and Medicine, U.S. Military ND.

emergence of pandemic influenza including overcrowding (with 100,000 soldiers in and out every day), live pigs and nearby live geese, duck and chicken markets, horses and 24 gases (some of them mutagenic) used in large 100-ton quantities to contaminate soldiers and the landscape. The final trigger for the ensuing pandemic was the return of millions of soldiers to their homelands around the entire world in the autumn of 1918 (see Oxford et al., 2005).

Why is so much attention paid to this nearly 100-year-old event? It is in part because of the high infection and kill rates as well as it being a vision of what might happen again if surveillance is downgraded. Many situations are in fact difficult to assess, especially with the origin and nature of zoonotic diseases. Public panic occurs, social distances increase and intolerance of others may grow. We point as examples to the Western responses to Avian flu (H5N1) which started spreading through Asia in 2003, swine flu (1918, 1976, 1988, 2009) as well as the unrelated SARS virus in 2003. All seemed to derive from environments in which animals and people live close together and appear to be insanitary events waiting to explode and spread disease and death globally. Furthermore, the 1976 swine flu led to wide spread vaccination, which led to significant numbers of neuro-muscular complaints and lawsuits against companies and levels of government. Cultural responses to preventive medical interventions have many cues.

Thus far, excluding the cancers, we have largely dealt with short latencies, still often made difficult to assess by confounding factors. Our final example of time lag is vCJD (variant Creutzfeldt-Jakob Disease), a rare, degenerative, fatal brain disorder in humans with a latency over 50 years or more. Although experience with this new disease is limited, evidence to date indicates that there has never been a case of vCJD transmitted through direct contact of one person with another (CDC, 2014). While there remain fewer than 250 cases worldwide, they are found in 12 countries, with about three quarters being in the UK. Virtually all cases have appeared in countries in which cattle are infected with BSE (bovine spongiform encephalopathy, or mad cow disease). The most serious outbreak of BSE occurred in 1992 in the UK, which reached its peak incidence in January 1993 at almost 1,000 new cases per week. The outbreak may have resulted from the feeding of scrapie-containing sheep meat-and-bone meal to cattle. There is strong evidence and general agreement that the outbreak was amplified by feeding rendered bovine meat-and-bone meal to young calves. At this time it was largely young people affected by vCJD, these possessing atypical clinical features, with prominent psychiatric or sensory symptoms. At the time of clinical presentation delayed onset of neurologic abnormalities included ataxia (uncoordinated movements) within weeks or months, dementia (decline in memory and thinking) and myoclonus (muscle twitches) late in the illness, a duration of illness of at least six months, and a diffusely abnormal non-diagnostic electroencephalogram. It may be that this group forms the canary in the coalmine, as vCJD will emerge as those who have eaten infected beef age. Infection in rare cases has been also between person to person, and some groups (Black British) seem to have a higher incidence rate than would be expected from the case data (see

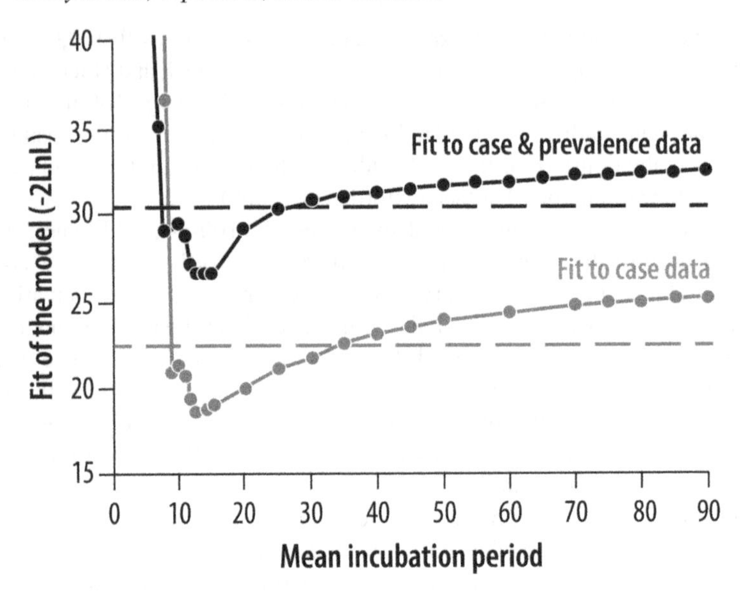

Figure 7.6 Likelihood profile for the mean incubation period of vCJD
Source: Ghani et al. (2003), under CC BY 4.0 from BMC Infectious Diseases.

Garske and Ghani, 2010). New cases are likely to be scarce, although a second peak may emerge over time. And the quite frequent reporting of BSE leads to public if not scientific concerns.

The longest latency period is intergenerational, but here we focus on conditions and responses, many of which have concerned reproductive and developmental outcomes. Although much depends on the speed with which contaminants are expelled from the body, numerous studies have shown that exposure to environmental contaminants during the foetal period or early childhood can cause long-term health effects such as increased susceptibility to cancer, damage to the immune and reproductive systems and common chronic diseases in adult age. Low-level exposures before or shortly after birth often produce more damaging and longer-lasting harm than exposures at higher levels in later childhood or adult life. For example, Patayová et al. (2013) related placental transfer, quantified by the cord-to-maternal-serum concentration ratio (C/M), of five organochlorine pesticides (OCP) hexachlorobenzene (HCB), β-hexachlorocyclohexane (β-HCH), γ-hexachlorocyclohexane (γ-HCH) , p,p'-DDT, p,p'-DDE and 15 polychlorinated biphenyl (PCB) congeners. The placenta is a weak barrier to OCP and PCB. In their study in Slovakia, Patayova et al. found that serum concentration was associated with gestational alcohol consumption, fewer illnesses during pregnancy, maternal age and maternal weight. Drinking can further lead to increased levels of several PCB congeners. But even these generational transfers are difficult to work out.

In terms of intergenerational effects of foetal PCB exposure, Forns et al. (2012) report that although negative effects on neuropsychological development have been observed in previous studies on PCB exposure, there are inconsistencies in these effects at current exposure levels, since for previous generations' exposure to these compounds was much lower. They aimed therefore to sort out the effects of prenatal and postnatal PCB exposure on neuropsychological development at 4 years of age. Individual congener analyses yielded significant detrimental effects of prenatal PCB153 for these developments, while no effects were reported for other congeners. The levels of PCBs at 4 years of age were not associated with neuropsychological development. Thus, prenatal exposure to low-level concentrations of PCBs, particularly PCB153, was associated with an overall deleterious effect on neuropsychological development at 4 years of age, including negative effects on executive function, verbal functions and visuo-spatial abilities, but not on motor development. It is not only maternal transfers that are a concern, but also paternal ones. Epidemiological studies have demonstrated that the exposure of fathers to various drugs, toxins and other chemicals, such as endocrine disruptors, before mating is associated with altered behavioural development in their children, even after accounting for other potential confounding lifestyle variables. Males exposed to such drugs and toxins as opiates, cyclophosphamide, ethylene dibromide and lead have been found to sire offspring with developmental and behavioural impairments, with these effects in several cases being transmissible via the male line to second and third generations (see Curley et al., 2011). Some of these insults depend upon a critical transfer point in foetal development. Yet environmental insults are one small piece in the puzzle.

Our worries about these possible but difficult-to-measure effects on children, especially if they are prenatal in impact, are great. As indicated in Chapter 9, environmental and health policy responses tend to favour controlling substances that selectively impact children. Virtually all cultures at all times have safeguarded their children for economic as well as emotional reasons. Children represent innocence, a gilded past and a commitment to the future. The unborn is a particularly salient image, and many therefore are concerned about what appear to be minor vagaries of everyday life – vaccination, low level of chemical exposure and so on. Concern seems to be everywhere as the diagrams from the World Hair Research (2010) show. Using data from Smrcka and Leznarova (1998), Guo et al. (2004) and Weiss et al. (2004), they chart the possible impacts on the unborn, such as lower birth weight, delayed cognitive development, heart defects and hyperpigmentation.

Psychosocial responses to latency

Long latency also means long-term exposure to a phenomenon which is likely to lead to psychosocial as well as physical consequences. Thoits (2010) summarizes this well, commenting that five dimensions have been identified. First, when stressors (negative events, chronic strains and traumas) are measured comprehensively, their damaging impacts on physical and mental health can be substantial. Second, differential exposure to stressful experiences is a way in which gender, racial-ethnic, marital status and social class inequalities in physical and mental health are produced and intensified. Third, minority group members can also suffer from discrimination stress. Fourth, stressors proliferate over the life course and across generations, widening health gaps between advantaged and disadvantaged group members. Fifth, the impacts of stressors on health and well-being may be reduced when persons have high levels of mastery, self-esteem and/or social support. It has been noted, for example, that long-term exposure to adverse environmental conditions in neighbourhoods can lead to poor health outcomes. These stressors continue to have impact despite policy interventions to tackle these situations. This is made complex by the patterning of residential environments by social class and race/ethnicity, factors which may contribute to the development and maintenance of health inequalities, especially through their interaction with other socially patterned individual-level characteristics including personal and family material resources, psychosocial supports and coping strategies (see Diez Roux and Mair, 2010). All these elements conspire to create higher levels of adverse psychosocial impact. Several studies have found that this wear and tear on body and mind (allostatic load) begins at an early age, among adolescents who are in most respects at the peak of their health (see Theall et al., 2012). The overlay with these neighbourhood effects is often long-term poverty. Evans and English (2002) note that one in five children growing up in poverty in America have elevated risk for socio-emotional difficulties that is further exacerbated by exposure to multiple physical and psychosocial stressors. Evans and English (2002) showed that 8- to 10-year-old, low-income, rural children confront a wider array of multiple physical (substandard housing, noise, crowding) and psychosocial (family turmoil, early childhood separation, community violence) stressors than do their middle-income counterparts. They replicate findings on self-reported distress among inner-city minority children.

Long-term exposure to stressful situations can worsen feelings of vulnerability (see Chapter 6). Responses to such stress are often internalized (and may well be latent) until some trigger releases negative reactions, such as property destruction, theft or violence. But what if this long-term exposure is removed? Ludwig et al. (2013) examine the long-term effects on low-income parents and children of moving from very disadvantaged to less distressed neighbourhoods in the Moving to Opportunity (MTO) demonstration. Via random lottery, MTO offered housing vouchers to families with children living in high-poverty public housing projects to move to less-distressed areas. MTO randomization generates large, persistent differences in neighbourhood conditions for otherwise comparable groups. They found that 10 to 15 years after randomization, MTO–assisted moves improve several key adult mental and physical health outcomes but have no consistent detectable impacts on adult economic self-sufficiency or children's educational achievement outcomes. Gender differences in the effects of MTO moves on youth risky behaviours and health were found, with girls doing better in some ways while boys do worse. Despite the mixed MTO impacts on the standard outcomes that have dominated the neighbourhood-effects literature, MTO moves generate a large gain in subjective well-being (SWB) for adults. Ending the latent impacts and long-term exposure therefore positively changes some things but not others.

Latency can complicate risk assessment further if a substance is essentially omnipresent. Long-term exposure to environmental pollutants with long latencies can often be invisible or considered part of everyday life. Thus impact on the skin from nickel (in cement), mercury (thermostats), arsenic (pressure-treated wood), chromium (paints) and VOCs (volatile organic compounds) in fumes from gasoline, paint, adhesives can affect other systems and organs through dermal absorption. Though VOCs and benzene can be eye and skin irritants, their main impacts are through inhalation. As WHO (2010) notes, benzene is commonplace coming from the processing of petroleum products, coking of coal and production of toluene, xylene and other aromatic compounds and is used in industrial and consumer products as a chemical intermediate and as a component of petrol (gasoline) and heating oils. The presence of benzene in petrol and as a widely used industrial solvent can result in significant occupational exposure and widespread emissions to the environment. Automobile exhaust accounts for the largest source of benzene in the general environment. Off-gassing from building materials and structural fires leads to increased atmospheric benzene levels. Industrial discharge, landfill leachate and disposal of benzene-containing waste are also sources of exposure. There is widespread recognition of the health impacts of benzene. It is a well-established cause of cancer in humans (IARC – Type 1 carcinogen). It causes acute myeloid leukaemia (acute non-lymphocytic leukaemia). There is limited evidence that benzene may also cause acute and chronic lymphocytic leukaemia, non-Hodgkin's lymphoma and multiple myeloma. Chronic exposure to benzene can also reduce the production of both red and white blood cells from bone marrow in humans, resulting in aplastic anaemia. Most of the studies on which these conclusions are based are for workers in industries that use benzene, but the impact of lower-level exposures for the rest of us is much more difficult to tease out (Krewski et al., 2000). There is public health recognition of its dangers, but exposure is a global problem and one, given the centrality of petroleum products, that is hard to tackle. There are exposure limits and the use of some less toxic substitutes, but production and exposure continue to increase in the Middle East and Africa. Benzene exposure has also increased and remains high in petroleum-producing provinces and states in North America. Oil sands mining in Alberta has had such an impact. Studying such exposure sites may be a useful step forward in further characterizing health impacts of lower-dose chronic exposure to benzene.

Politics of delay

Bureaucracy is often slow, and there seems to be no exception in the case of environmental regulation to mitigate against long-term exposure, long or uncertain latencies and the resultant health impacts. But interests – government itself, industry and the public – can delay actions. As we have seen, the Canadian government has delayed action on asbestos, greenhouse gas emissions and climate change regulations more broadly. Political context, including the nature of the state (centralized, federal), the role of bureaucracy (politicized, independent), international agreements, the connection between political and economic interests and government response to public fears (immediate or anticipatory) will affect speed and type of response to latent environment and health threats. Furthermore there is often interaction between these contextual factors themselves as well as with scientific assessments of the benefits of regulation. In a study of more than 100 U.S. department decisions (e.g., health transportation security, housing) Shapiro and Morrall (2012) found scientific assessments were not greatly used. They were, however, of greatest benefit when there was little public comment and if regulations were issued well before the end of a government's term. Overall prevention is difficult, because

> the success of prevention is invisible, lacks drama, often requires persistent behavior change, and may be long delayed; statistical lives have little emotional effect, and benefits often do not accrue to the payer; avoidable harm is accepted as normal, preventive advice may be inconsistent, and bias against errors of commission may deter action; prevention is expected to produce a net financial return, whereas treatment is expected only to be worth its cost; and commercial interests as well as personal, religious, or cultural beliefs may conflict with disease prevention.
>
> (Fineberg, 2013, p. 85)

For example, Outka (2012) demonstrates how regulations and laws protect fossil fuel exploration and production in the United States, meaning that their costs are internalized and literally sent into the air to the disadvantage of health, responses to climate change and the development of renewable energies.

The power of tobacco companies to influence public and political opinion and decision making is well known (see Weishaar et al., 2012). More recently, there has been some investigation of 'Big Food' and its possible impact on chronic diseases in the developed and, through foreign direct investment, the developing world. Stuckler and Nestle (2012) comment, "Big Food attains profit by expanding markets to reach more people, increasing people's sense of hunger so that they buy more food, and increasing profit margins through encouraging consumption of products with higher price/cost surpluses" (see http://journals.plos.org/plosmedicine/article?id=10.1371/journal.pmed.1001242). Brownell and Warner (2009, p. 259) point to significant parallels between tobacco and food industries. They comment that the tobacco industry had a playbook, a script, that emphasized personal responsibility, paying scientists who delivered research that instilled doubt, criticizing the 'junk' science that found harms associated with smoking, making self-regulatory pledges, lobbying with massive resources to stifle government action, introducing 'safer' products and simultaneously manipulating and denying both the addictive nature of their products and their marketing to children. The script of the food industry is both similar to and different from the tobacco industry script, as the food one is varied, with thousands of producers. Koplan and Brownell (2010) see the food industry reframing issues, such as emphasizing 'balance' and 'calories out' (exercise). Since these companies are in the 'calories in' business, focusing attention on lack of physical activity may divert attention from food. The industry argues that there are no bad foods and that only the totality of the diet counts. Health experts agree widely that population consumption of some foods (e.g., sweetened beverages and fast foods) should decrease. As with fossil fuels, much of the cost of obesity-related illness is paid for by public funds – the rest privately. Little of the cost is paid for by the food companies, who may simultaneously see themselves as supporting the virtues of personal freedom and condemning the nanny state. These arguments appear, at present, to gel with the public and therefore political views – signalling little need for action, as people must be free to eat (and drive) as they please. State intervention may be feared more than the risks of benzene exposure and obesity-related illnesses. Where personal choice is at stake for a mass of the population, risk is nowhere, and falling ill is an individual responsibility. But the ecological and population risks and impacts cannot be ignored. These remain latent problems for the future. But associations are not always easy to see, as the connection between fast food restaurants and obesity rates show. Proximity does seem to matter. Changes in the supply of fast food restaurants matter. Among ninth graders, a fast food restaurant within 0.1 miles of a school results in a 5.2% increase in obesity rates. Among pregnant women, a fast-food restaurant within 0.5 miles of residence results in a 1.6% increase in the probability of gaining more than 20 kilos (see Currie et al., 2010). Small associations are present, but large sample sizes and the use of proxy variables may not persuade those valuing personal choice.

In this chapter, we have been mainly concerned with the delay that often occurs between an environmental exposure and a health consequence. With infectious diseases, biological forces may be responsible for the delay, but in other circumstances the delay can be caused or exacerbated by individual, societal, political and legal actions. Thus the latency of the impact of being exposed to asbestos, mercury, PCBs and so on is clouded by individual actions and needs, for example employment, benefits of technology (Chapter 6), and also by company approaches which may deny wrongdoing and present lopsided evidence. Governments can support company positions and may delay regulation unless or until public reaction demands remedial activity or regulation.

References

Barry, J.M. 2005. '1 The Story of Influenza: 1918 Revisited: Lessons and Suggestions for Further Inquiry.' in Knobler, S.L., Mack, A., Mahmoud, A., & Lemon, S.M. (eds.). *The Threat of Pandemic Influenza: Are We Ready? Workshop Summary (2005)*. Washington, DC: The National Academies Press. pp 60–61.

Bartrip, P. 1998. Too little, too late? The home office and the asbestos industry regulations, 1931. *Medical History*, 42(4), 421–438.

Bell, M.L., & Davis, D.L. 2001. Reassessment of the lethal London fog of 1952: Novel indicators of acute and chronic consequences of acute exposure to air pollution. *Environmental Health Perspectives*, 109(Suppl 3), 389.

Braga, A.L.F., Zanobetti, A., & Schwartz, J. 2001. The lag structure between particulate air pollution and respiratory and cardiovascular deaths in 10 US cities. *Journal of Occupational and Environmental Medicine*, 43(11), 927–933.

Brownell, K.D., & Warner, K.E. 2009. The perils of ignoring history: Big Tobacco played dirty and millions died. How similar is Big Food? *Milbank Quarterly*, 87(1), 259–294.

Burnett, R.T., Brook, J.R., Yung, W.T., Dales, R.E., & Krewski, D. 1997. Association between ozone and hospitalization for respiratory diseases in 16 Canadian cities. *Environmental Research*, 72(1), 24–31.

Canada Is On the Sidelines When It Comes to Banning Asbestos Trade. 2015. *The Toronto Star*. Available at: http://www.thestar.com/opinion/commentary/2015/02/27/canada-is-on-the-sidelines-when-it-comes-to-banning-the-asbestos-trade.html (Accessed 24 March 2015).

CBC News. 2013. *Quebec's Asbestos Promotion Policy May Be Ending*. Available at: http://www.cbc.ca/news/canada/montreal/quebec-s-asbestos-promotion-policy-may-be-ending-1.1341364 (Accessed 24 March 2015).

CDC. 2014. *vCJD (Variant Creutzfeldt-Jakob Disease)*. Available at: http://www.cdc.gov/ncidod/dvrd/vcjd/factsheet_nvcjd.htm (Accessed 24 March 2015).

Clapp, R., Howe, G., & Jacobs Lefevre, M. 2005. *Environmental and Occupational Causes of Cancer*. Available at: http://www.sustainableproduction.org/downloads/Causes%20of%20Cancer.pdf (Accessed 20 March 2015).

Curley, J.P., Mashoodh, R., & Champagne, F.A. 2011. Epigenetics and the origins of paternal effects. *Hormones and Behavior*, 59(3), 306–314.

Currie, J., DellaVigna, S., Moretti, E., & Pathania, V. 2010. The effect of fast food restaurants on obesity and weight gain. *American Economic Journal: Economic Policy*, 2, 32–63.

Diez Roux, A.V., & Mair, C. 2010. Neighborhoods and health. *Annals of the New York Academy of Sciences*, 1186(1), 125–145.

Engleson, G., & Herner, T. 1952. Alkyl mercury poisoning. *Acta Paediatrica*, 41, 289–294.

Evans, G.W., & English, K. 2002. The environment of poverty: Multiple stressor exposure, psychophysiological stress, and socioemotional adjustment. *Child Development*, 73(4), 1238–1248.

Faith, F. 2013. *Study Revisits Health Risk of Chrysotile: Why Is This Still a Debate in 2013?* Available at: http://www.asbestos.com/news/2013/02/01/health-risk-of-chrysotile/ (Accessed 24 March 2015).

Fineberg, H.V. 2013. The paradox of disease prevention: Celebrated in principle, resisted in practice. *Journal of the American Medical Association*, 310(1), 85–90.

Forns, J., Torrent, M., Garcia-Esteban, R., Grellier, J., Gascon, M., Julvez, J., . . . & Sunyer, J. 2012. Prenatal exposure to polychlorinated biphenyls and child neuropsychological development in 4-year-olds: An analysis per congener and specific cognitive domain. *Science of the Total Environment*, 432, 338–343.

Garske, T., & Ghani, A.C. 2010. Uncertainty in the tail of the variant Creutzfeldt-Jakob disease epidemic in the UK. *PLoS One*, 5(12), e15626. doi: 10.1371/journal.pone0015626.

Gee, D., & Greenberg, M. 2002. 'Asbestos: From "Magic" to Malevolent Mineral.' in European Environmental Agency. *Late Lessons from Early Warnings: The Precautionary Principle 1896–2000*. Copenhagen: EEA. pp 52–63.

Ghani, A.C., Donnelly, C.A., Ferguson, N.M., & Anderson, R.M. 2003. Updated projections of future vCJD deaths in the UK. *BMC Infectious Diseases*, 3(1), 4.

Grandjean, P., Satoh, H., Murata, K., & Eto, K. 2010. Adverse effects of methylmercury: Environmental health research implications. *Environmental Health Perspectives*, 118(8), 1137–1145.

Guo, Y.L., Lambert, G.H., Hsu, C.C., & Hsu, M.M. 2004. Yucheng: Health effects of prenatal exposure to polychlorinated biphenyls and dibenzofurans. *International Archives of Occupational and Environmental Health*, 77(3), 153–158.

Haber, S.E., & Haber, J.M. 2011. Malignant mesothelioma: A clinical study of 238 cases. *Industrial Health*, 49(2), 166–172.

Harada, M. 2004. *Minamata Disease*. Tokyo: Kumamoto Nichinichi Shinbun Culture Centre. (First published in Japanese 1972.)

Ishimure, M. 1990. *Paradise in the Sea of Sorrow*. Tokyo: Japan Publishing.

Kjellstrom, T., Kennedy, P., Wallis, S., & Mantell, C. 1986. *Physical and Mental Development of Children with Prenatal Exposure to Mercury from Fish. Stage I: Preliminary Tests at Age 4*. Report 3080. Solna, Sweden: National Swedish Environmental Protection Board.

Koplan, J.P., & Brownell, K.D. 2010. Response of the food and beverage industry to the obesity threat. *Journal of the American Medical Association*, 304(13), 1487–1488.

Krewski, D., Snyder, R., Beatty, P., Granville, G., Meek, B., & Sonawane, B. 2000. Assessing the health risks of benzene: A report on the benzene state-of-the-science workshop. *Journal of Toxicology and Environmental Health Part A*, 61(5–6), 307–338.

Lee, H.R., Hwang, K.A., Nam, K.H., Kim, H.C., & Choi, K.C. 2014. Progression of breast cancer cells was enhanced by endocrine-disrupting chemicals, triclosan and octylphenol, via an estrogen receptor-dependent signaling pathway in cellular and mouse xenograft models. *Chemical Research in Toxicology*, 27(5), 834–842.

Letašiová, S., Medve'Ova, A., Šovčíková, A., Dušinská, M., Volkovová, K., Mosoiu, C., & Bartonová, A. 2012. Bladder cancer, a review of the environmental risk factors. *Environmental Health*, 11(Suppl 1), S1–S11.

Ludwig, J., Duncan, G.J., Gennetian, L.A., Katz, L.F., Kessler, R.C., Kling, J.R., & Sanbonmatsu, L. 2013. Long-term neighborhood effects on low-income families: Evidence from moving to opportunity. *American Economic Review*, 103(3), 226–231.

Madkour, M.T., El Bokhary, M.S., Awad Allah, H.I., Awad, A.A., & Mahmoud, H.F. 2009. Environmental exposure to asbestos and the exposure–response relationship with mesothelioma. *Eastern Mediterranean Health Journal*, 15(1), 25–38.

Mayor of London. 2002. *50 Years On*. Available at: http://legacy.london.gov.uk/mayor/environment/air_quality/docs/50_years_on.pdf (Accessed 21 March 2015).

New Minamata Compensation. 2010. *The Japan Times*. Available at: http://www.japantimes.co.jp/opinion/2010/04/05/editorials/new-minamata-compensation/#.VdTAp_lVhBc (Accessed 24 March 2015).

Outka, U. 2012. Environmental law and fossil fuels: Barriers to renewable energy. *Vanderbilt Law Review*, 65(6), 1679–1721.

Oxford, J.S., Lambkin, R., Sefton, A., Daniels, R., Elliot, A., Brown, R., & Gill, D. 2005. A hypothesis: The conjunction of soldiers, gas, pigs, ducks, geese and horses in Northern France during the Great War provided the conditions for the emergence of the "Spanish" influenza pandemic of 1918–1919. *Vaccine*, 23(7), 940–945.

Papania, M.J., Wallace, G.S., Rota, P.A., Icenogle, J.P., Fiebelkorn, A.P., Armstrong, G.L., . . . & Seward, J.F. 2014. Elimination of endemic measles, rubella, and congenital rubella syndrome from the Western hemisphere: The US experience. *JAMA Pediatrics*, 168(2), 148–155.

Patayová, H., Wimmerová, S., Lancz, K., Palkovičová, Ľ., Drobná, B., Fabišiková, A., . . . & Trnovec, T. 2013. Anthropometric, socioeconomic, and maternal health determinants of placental transfer of organochlorine compounds. *Environmental Science and Pollution Research*, 20(12), 8557–8566.

Patterson, K.D., & Pyle, G.F. 1991. The geography and mortality of the 1918 influenza pandemic. *Bulletin of the History of Medicine*, 65(1), 4–21.

Potter, C.W. 2001. A history of influenza. *Journal of Applied Microbiology*, 91(4), 572–579.

RAND. 2014. *Investigating Time Lags and Attribution in the Translation of Cancer Research*. Available at: http://www.rand.org/content/dam/rand/pubs/research_reports/RR600/RR627/RAND_RR627.pdf (Accessed 20 March 2015).

Roselli, M. 2014. *The Asbestos Lie: The Past and Present of an Industrial Catastrophe*. Brussels: European Trade Union Institute.

Shah, A.S., & Newby, D.E. 2014. Less clarity as the fog begins to lift. *Heart*, 100(14), 1073–1074. doi: 10.1136/heartjnl-2014-305877.

Shapiro, S., & Morrall III, J.F. 2012. The triumph of regulatory politics: Benefit–cost analysis and political salience. *Regulation & Governance*, 6(2), 189–206.

Smrcka, V., & Leznarova, D. 1998. Environmental pollution and the occurrence of congenital defects in a 15-year period in a south Moravian district. *Acta Chirurgiae Plasticae*, 40(4), 112–114.

Spyker, J.M., Sparber, S.B., & Goldberg, A.M. 1972. Subtle consequences of methylmercury exposure: Behavioral deviations in offspring of treated mothers. *Science*, 177, 621–623.

Stuckler, D., & Nestle, M. 2012. Big food, food systems, and global health. *PLoS Medicine*, 9(6), 678.

Theall, K.P., Drury, S.S., & Shirtcliff, E.A. 2012. Cumulative neighborhood risk of psychosocial stress and allostatic load in adolescents. *American Journal of Epidemiology*, 176(Suppl 7), S164–S174.

Thoits, P.A. 2010. Stress and health major findings and policy implications. *Journal of Health and Social Behavior*, 51(Suppl 1), S41–S53.

Thurston, G.D., Ito, K., Lippmann, M., & Hayes, C. 1989. Reexamination of London, England, mortality in relation to exposure to acidic aerosols during 1963–1972 winters. *Environmental Health Perspectives*, 79, 73–82.

U.S. Army photographer (public domain). Available at www.army.mil/-images/2008/09/24/22729/army.mil-2008-09-25-103608.jpg.

Weishaar, H., Collin, J., Smith, K., Grüning, T., Mandal, S., & Gilmore, A. 2012. Global health governance and the commercial sector: A documentary analysis of tobacco company strategies to influence the WHO framework convention on tobacco control. *PLoS Medicine*, 9(6), 703.

Weiss, B., Amler, S., & Amler, R.W. 2004. Pesticides. *Pediatrics*, 113(Suppl 4), 1030–1036.

WHO. 2010. *Exposure to Benzene*. Available at: http://www.who.int/ipcs/features/benzene.pdf (Accessed 23 March 2015).

WHO. 2014. *World Health Organization, Measles – Fact Sheet N°286*. Geneva: Author.

8 Who suffers most and what can be done?

Introduction

Pollution and contamination threats as well as health itself are unevenly distributed within society temporally (Chapter 7), spatially and through the social hierarchy – all key reasons that social scientists are interested in environment and health problems. In many cases vulnerable groups in society (e.g., low income, visible minorities and lower social class) bear a disproportionate burden of the negative aspects of industrial development. In general, environmental justice research in the social sciences asks where and when these vulnerable groups are more exposed to pollution and (potentially) polluting facilities and why. Research concerns range from the spatial coincidence of pollution and social groups to questions about the distribution and exercise of power among these groups in relation to environmental hazards. What is the percentage of visible minorities living in close proximity to sites that handle and release potentially harmful toxins compared to the percentage living farther away? Is there uneven access to decision making that helps explain disproportionate exposure to pollution threats? How do unfair spatial and temporal pollution relationships emerge? These are some of the key questions for environmental justice to which questions about health specifically may be added. What is the health status in communities that are vulnerable and disproportionately exposed? How are the toxins/pollutants implicated in unequal health outcomes, and how are the presence and aesthetics (e.g., sight, smell, sound) of a site/facility also implicated in health outcomes?

This chapter links together two major literatures: environmental justice and health inequalities, with an emphasis on the former since it is largely disproportionate exposure that is a main concern here. That is, though exposure to pollution and hazardous facilities may certainly be implicated in exacerbating health inequalities in various ways, disproportionate exposure itself, regardless of health outcomes, warrants justice considerations as well. The chapter begins with a discussion of health inequalities, followed by a more detailed discussion of environmental justice research and highlights of how these literatures intersect. Along the way key methodological issues are discussed as well as some broader philosophical questions about current directions in environmental justice research.

Health inequalities

There is ongoing concern about the distribution of health within society. This phenomenon is known variously as the problem of inequalities in health (Marmot, 2001), health inequities (Marmot et al., 1991), race and health (Williams et al., 1994) and health injustice (Venkatapuram, 2012). These terms all refer to the idea that ill health and, ultimately, premature death happen more among social groups lower in the social hierarchy. There are numerous studies which compare individuals from large datasets to determine the relationship between health and social class (e.g., Marmot et al., 1991; Ross et al., 2000). The evidence is rather overwhelming that higher social classes do better and lower social classes do not do nearly as well in terms of health – whether poor health is measured as the number of chronic symptoms (Gee, 2002; Huurre et al., 2003), self-assessed health status (van Doorslaer et al., 1997) or age at death (Marmot, 2005). The usual challenges of measuring social class apply (Krieger et al., 1997), but in all cases health status increases as social class increases, even in places with publicly funded health care systems like Canada (Raphael, 2000). Social class has been measured as (median) income (Kaplan et al., 1996; Wilkinson, 1997; Subramanian and Kawachi, 2004), occupational class (Marmot et al., 1991) and even with composite indexes that account for the multi-dimensionality of class like household, neighbourhood, and temporal influences (Wagstaff et al., 1991; Krieger et al., 1997).

What is most dramatic is that with few exceptions, no matter how social class and health are measured there tends to be a consistent social gradient of health. For example, dividing the social categories in different ways (e.g., five occupational classes; six income categories) does not change the fact that health generally declines moving from each higher-status group to the next lower one. Though this is not a perfect gradient for women in Figure 8.1, from the well-known Whitehall II study in the United Kingdom, in all but two cases health measured as a 36-item "physical functioning" index declines moving down the occupational class hierarchy (Marmot and Brunner, 2005).

Further, health is unevenly distributed among racial and ethnic groups in society (Rogers 1992; Nazroo 2003). For example, Navarro (1990) highlighted that African Americans had an average age at death of 70 years compared to 76 for white Americans. The gap had improved by about two years in 2011 with age at death figures at 75.3 and 78.8 respectively (CDC, 2015), but inequalities persist. Navarro emphasizes that despite the fact that statistically, there are separate and independent effects of both race and social class (income), they are also synergistic whereby race and income likely have a mutually reinforcing effect on premature death. Likewise Krieger (2012) recommends moving away from 'race' as a monolithic, reified category in epidemiologic studies of health inequalities towards a more socioecological approach. She suggests that three aspects of race should be considered: structural issues (e.g., institutional racism), individual issues (e.g., everyday social interactions) and the impact of research itself, since she claims that many studies have underestimated of the impact of racial discrimination on health. Further in the chapter the same principle of 'the whole effect is

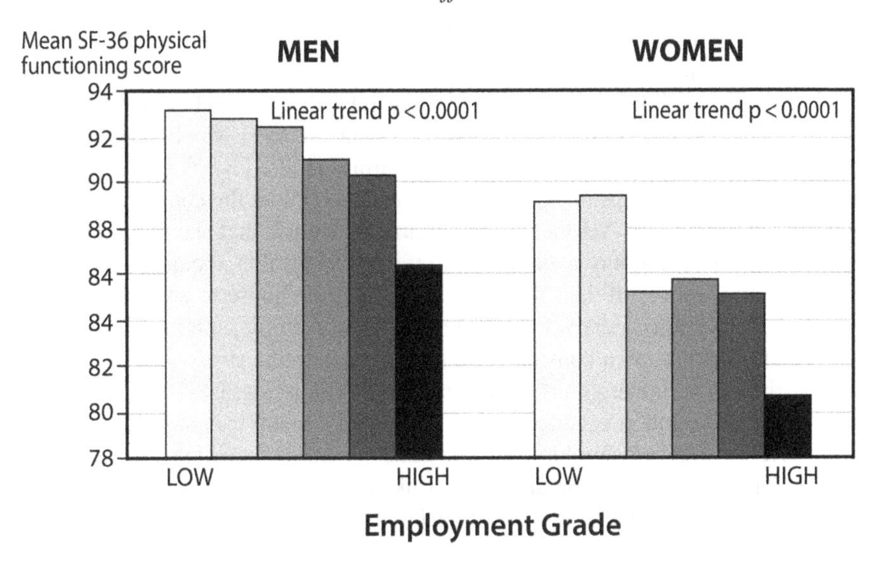

Figure 8.1 Inequalities in health in the UK: SF-36 physical functioning score by employment grade, men and women aged 40–59 years

Source: Marmot and Brunner (2005), reproduced with kind permission from Oxford University Press.

greater than the sum of the parts' also applies to conceptualizing environmental justice research.

One of the central threads to the health inequalities literature is the quest for underlying causes for the health differences by social class race and ethnicity. Some suggest that material deprivation is the principle reason for the social health gradient. This makes intuitive sense in health care systems where access to health care is determined by the personal ability to pay. Yet the health inequalities persist in places like the UK and particularly Canada, where access to health care is provided through collective funding, and money is theoretically not a barrier to access. Another theory is that it is not material deprivation per se that prevents access, but it is the interplay between relative deprivation and psychosocial stress. This may be considered a 'keeping up with the Joneses' thesis, in the sense that knowing you do not have the same material circumstances as your wealthier neighbour puts tacit pressure on you to improve your social circumstances. Thus, having access to enough material goods to survive (needs) may not be as important as having enough material goods to fulfil social desires (wants). This idea links with theories of social stress in the sense that those whose stress is due to the inability to realize their material 'wants' experience more serious deterioration than those whose stress comes from a position of their own choosing (Marmor et al., 1994). This likely has something to do with the perceived amount of control one has over their life and work (van Dijk et al., 2013) – where, for example, white-collar individuals may perceive that they have high control, whereas blue-collar workers perceive low control (Shippee et al., 2012).

A more controversial explanation for the underlying pattern in health inequalities is that income and other measures of social class are actually proxies for lack of knowledge (Gottfredson, 2004). This is consistent with behaviour theories in public health – for example, information deficit theory – which suggest that risky personal health behaviours (e.g., poor eating habits, low exercise, unprotected sex) are due to gaps in exposure to information about the consequences of these riskier behaviours. Yet there are a number of studies that show that people experiencing ill health may actually be quite knowledgeable about the underlying behavioural causes of their maladies, but they still choose to engage in such behaviours for various socially rational reasons. For example, smoking and alcohol consumption are often considered to be antidotes to the stresses of everyday life, but they nevertheless cause numerous and serious diseases – particularly where high consumption is concerned. What is problematic from a public health point of view is that both smoking and alcohol consumption are highly socialized, whereby for example smoke breaks and a drink after work become routine aspects of daily work life, and breaking free of such habits leads to new risks like social isolation. In this sense such behaviours are highly engrained into localized social structures such that consumption and materiality (job, wealth) may be connected to somatic (addiction) and social (deprivation) concerns (Siegrist and Marmot, 2004; Laaksonen et al., 2005).

Environmental causes of health inequalities

Another growing theory of the underlying causes of inequalities in health is that differences across the social hierarchy are due to parallel differences in environmental exposure. Geographers and sociologists have been exploring this idea implicitly in a series of neighbourhood effects studies, in which neighbourhood is isolated as a variable in multi-level modelling to predict health outcomes while simultaneously controlling for other variables that might otherwise explain health inequalities, including income, occupational status and smoking behaviour (e.g., Macintyre et al., 2002; Sampson et al., 2002). That literature has tended to show that indeed neighbourhoods ('contextual effects') consistently have a significant influence on health over and above individual characteristics like income ('compositional effects'). Though these contextual/neighbourhood/environmental variables significantly predict health inequalities, they also tend to have relatively small effects compared to individual/compositional variables (Pickett and Pearl, 2001).

Yet such neighbourhood studies are ecological, focussed on wealth and social class at the neighbourhood level, and tend not to isolate pollution, toxins or facilities in the models. That is they do not *specifically* explore the idea that poorer communities in particular are exposed to more pollution and toxins than their wealthier counterparts; yet pollution exposures are sometimes implicated in the interpretation where neighbourhood effects are indeed significant.

Environmental exposures may also be positive, and similar patterns of injustice may be seen regarding positive features like green space. These may likewise be unevenly distributed up the social hierarchy whereby the wealthy have greater access to not only clean air and water but also protected and other natural areas. For example, in a study of England Mitchell and Popham (2008) show that people at the same income level are less likely to die from all causes of death and diseases of the circulatory system (but not from lung cancer or suicide) when they have greater exposure to green space. This suggests a protective effect for green space and is the topic of the growing 'therapeutic landscapes' literature (Gesler, 1992; Williams, 2007) discussed in Chapter 10. Likewise access to such amenities may increase resilience and the ability to cope with environmental threats as discussed in Chapter 6.

Health inequalities from environmental inequalities?

Before turning to the environmental justice literature in detail it is worth noting that the connections between that literature and the health inequalities literature are often implicit (Brulle and Pellow, 2006; Masuda et al., 2010; Wakefield and Baxter, 2010). That is, a key concern of the environmental justice movement is that pollution, contamination and facilities suspected to be releasing toxic substances all threaten health. Yet environmental justice studies tend to focus on pollution exposure and often do not specifically include health measures (Bowen, 2002). Likewise, health inequalities studies suggest environmental pollution may be a source of inequalities – for example, in neighbourhood-based studies – but they do not include any direct measures of exposure beyond the neighbourhood itself. There are a number of reasons the links between these two literatures are not yet that extensive.

One key difference is conceptual and political in terms of the nature of end goals. While for health the end goal may be equality – everyone should be the same – this may not be the case for environmental pollution and hazardous facility exposure. There is a tacit assumption that we should all at least have access to the resources that help maintain good health, yet the same is not necessarily true for environmental exposure in the sense that the latter may logically be uneven based on access to resources. That is, it may reasonably be argued that those who consume more resources should be exposed more to the pollution that is the consequence of their high consumption. This could theoretically curb consumption; that is, high resource consumers must also be proportionally exposed to the manufacturing facilities, waste sites and energy production facilities that produce the goods they consume. Though such an idea may be politically distasteful, the underlying basis is supported by ecological footprint analysis, whereby the goods and services one consumes may be converted to a land area required to produce those goods and services. Such analyses are used as comparative tools to raise awareness of the very high relative impact the wealthy have on resource use and waste production compared to their relatively poor neighbours (Wackernagel and Rees, 1996; Galli et al., 2012). In this sense, environmental exposure may be thought of in terms of inequities – an even distribution of exposure does not need to be the goal as long as high differentials in consumption patterns persist or worsen in society. Yet the burden of high exposure tends to remain on the most rather than least disadvantaged. An equitable principle for distributive justice is that those who consume least should be exposed to pollution least so they have one less burden. Health tends to be viewed as a fundamental right and is thus viewed more as something that should be equally distributed – everybody in society should be allowed to achieve the highest levels of health as freedom from disease and overall well-being. Regardless of whether *equity* in pollution exposure or *equality* of health are politically feasible at any scale in a globalized economy these are conceptual differences that influence how researchers think about the relationship between environmental injustice and health inequalities.

A second reason that the health inequalities studies do not include more detailed environmental exposure data is that such measurement is often beyond the scope of such studies. For decades demonstrating trends in health inequalities and their robustness over time and space has been a primary goal (Bleich et al., 2012; Mackenbach, 2012). For example, Bleich et al. (2012) show that health inequalities have decreased in the U.S. for life expectancy but increased in the UK over time at the same time the authors call for more rigorous methods for addressing health inequality policy goals. Mackenbach speculates on why reducing health inequalities has been relatively unsuccessful globally, including explanations such as consumption and mobility patterns and use of material resources, while environmental exposures are conspicuously absent from the list. The following sections detail how measuring pollution exposure is an equally complex task by tracing the evolution of environmental justice research.

Environmental justice

Do the poor and racial minorities tend to live in polluted environments more so than their wealthy white counterparts? What social processes lead to such patterns of injustice where they do exist? Modern-day momentum in the environmental justice movement is often traced to the early 1980s in the United States when the rural community of Afton in Warren County, North Carolina, a predominantly (84%) African American community was selected to be the site for a major PCB landfill. This not only sparked massive local protests, it garnered national attention for the issue of environmental justice. Environmental justice issues arise in more urbanized environments as well, as in the case of Camden, New Jersey, an example that may be used to sketch some of the key issues in environmental justice.

Camden, New Jersey: A case example

Camden is a modest-sized city comprised largely of relatively poor racial minorities (91% are persons of colour and 50% are below the poverty line) across the Delaware River from Philadelphia. Located in the so-called Rust Belt, it has historically been a site for large-scale manufacturing including chemical companies, a petroleum coke transfer station, waste incineration and sewage treatment and has been the site of no less than two Superfund[1] sites. This is a place that appeared in a *Rolling Stone* article under the title "Apocalypse, New Jersey: A Dispatch from America's Most Desperate Town" (Taibbi, 2011). In response to mounting concerns about the health impacts of the cumulative effects of these sites, a group of concerned residents formed the South Camden Citizens in Action (SCCA) in 1997 – a group with a firm base in the local church communities. The two main targets for their actions were St. Lawrence Cement and, due to perceive lack of inaction on pollution generally, the State of New Jersey's Department of Environmental Protection. SCCA's goal was to prevent environmental permits from being issued on the basis of cumulative effects and environmental racism. They launched a civil suit, citing the powerful Comprehensive Environmental Response, Compensation, and Liability Act (CERCLA). The state court decided in their favour and granted an injunction to prevent St. Lawrence Cement from operating under those particular permits. Soon after though, the United States Supreme Court clarified that individual citizens, including the SCCA, could not be plaintiffs according to the relevant legislation (Foster, 2004). Further, the courts decided that claims of environmental racism must show intent of defendants to harm minority populations. This decision has been criticized by the SCCA for setting an unachievable legal standard by requiring the legal equivalent of a 'smoking gun' – for example, a memo suggesting that pollution permits are being requested in Camden because the local African American and Hispanic populations are unlikely to fight against such permits – where none is ever likely to exist. Thus, while the community cannot easily demonstrate legally that this single development or a single permit for that development should not be granted, the pattern of cumulative effects of pollution from multiple sources in the community remains. Similar scenarios are repeated elsewhere as in Sarnia, Canada – a.k.a. 'Chemical Valley' – where the cumulative effects of a wide array of chemical industries are suspected to cause a range of negative health outcomes including a skewed gender ratio (more than expected in females) in the local First Nations community (Hoover et al., 2012; Atari et al., 2013). However, as other parts of this volume attest, under current justice systems it is very difficult to make convincing links to any single company or even group of companies and their emissions (Atari et al., 2011; Wiebe, 2012). Such situations have the potential to further reinforce a sense of powerlessness in a community already faced with the challenges of multiple disadvantages which can erode psychosocial well-being.

Conceptualizing environmental justice

As the Camden case suggests environmental justice at its core is about fairness in decision making and the distribution of potential environmental harms. Yet there are a host of different ways that researchers and activists have conceptualized this idea. An often-cited definition comes from the U.S. Environmental Protection Agency, since President Clinton put forward an executive order (12898) in 1994 insisting that "environmental justice in minority and low-income populations be duly considered in any government department action". The EPA definition is as follows:

> Environmental Justice is the fair treatment and meaningful involvement of all people regardless of race, color, national origin, or income with respect to the development, implementation, and enforcement of environmental laws, regulations, and policies. EPA has this goal for all communities and persons across this Nation. It will be achieved when everyone enjoys the same degree of protection from environmental and health hazards and equal access to the decision-making process to have a healthy environment in which to live, learn, and work.

Firmly embedded in this definition are issues of legal institutionalization of fairness not only in the distribution of environmental harms but with a particular emphasis on processes like access to decision making. Ironically, the academic research that received the early attention from policy makers and other academics did not so much concern *process* as it did the spatial *patterns* of exposure. This difference between the emphasis in the EPA definition and on-the-ground research is conceptualized in the literature as the distinction between *procedural equity* in the case of the former and *outcome equity* in the case of the latter (Cutter, 1995). That is, procedural equity or justice refers to equal access by individuals and groups to the power structures that lead to the distribution of environmental harms and benefits. Thus, outcome equity refers to the principle of distributive justice discussed above, while procedural equity refers to the principles of *due process* (e.g., due notice, the right to grieve, the ability to appeal) and *natural justice* (freedom from bias).

Environmental justice and environmental equity may also be differentiated in the sense that environmental justice is tightly linked to the historically rooted nature of a whole range of inequities in society – for example colonialism, slavery, racism and classism (Cutter, 1995; Pulido, 1996). Narrowing the framing to environmental inequity alone may be considered a subtle tactic for diverting attention from these interconnected social problems. In terms of linkages with the inequalities in health literature, they both concern unfair treatment combined with disproportionate exposure (Davies, 2013). In this way ideas about procedural justice are parallel to notions of justice in health. For example, equality may be used as a guiding principle for both universal access to health care and equal access to power in decision making over hazardous facility location. Thus, environmental justice can cover a whole range of issues including access to green space (Sister et al. 2010; Walker 2009), access to treaty rights (Silvern, 1999; Taylor, 2000; Weestra 2012), injustice at political sites (Dunion and Scandrett, 2003; Scholsberg, 2004), international environmental injustice (Swyngedouw and Heynen, 2003) and differences of gender, age and the rights of future generations (Dobson 1998; Buckingham-Hatfield et al., 2005), as well as issues surrounding political, cultural and economic environments (Agyeman et al., 2003). This has been one of the challenges for the environmental justice movement; it is an umbrella term and means very different things depending on the context in which it is used. Many issues that were in previous generations considered simply social justice – race, cultural, poverty or civil rights issues – may now reasonably be reframed as environmental justice issues without too much effort. Though the term may suggest that the environment has taken centre stage, indeed the emphasis still tends to remain on how goods, bads and power are distributed in society.

Outcome/spatial/distributive environmental equity and justice

These next two sections describes some of the research and debates that have emerged within outcome and procedural justice research, the two very broad and interconnected areas of environmental justice enquiry defined earlier. Many of the landmark environmental justice studies concern the spatial distribution or patterns of environmental threats and vulnerable social groups in the United States – the 'outcomes' of decision making. Perhaps the two most influential outcome equity studies were the U.S. Government Accounting Office (USGAO; 1983) and United Church of Christ (UCC; 1987) studies. The USGAO study looked at the racial composition in communities around four of the largest U.S. hazardous waste sites in the South and found that three of the four sites had between 52% and 90% African American residents – disproportions that pointed clearly towards environmental injustice. The more extensive UCC study looked at 455 commercial hazardous waste facilities in the U.S. and assessed the percentage of racial minorities and low-income status as well as dwelling value by zip code unit. Figure 8.2 shows that as exposure to zip codes with hazardous waste increased so too did the percentage of minority residents (African Americans, Hispanics and

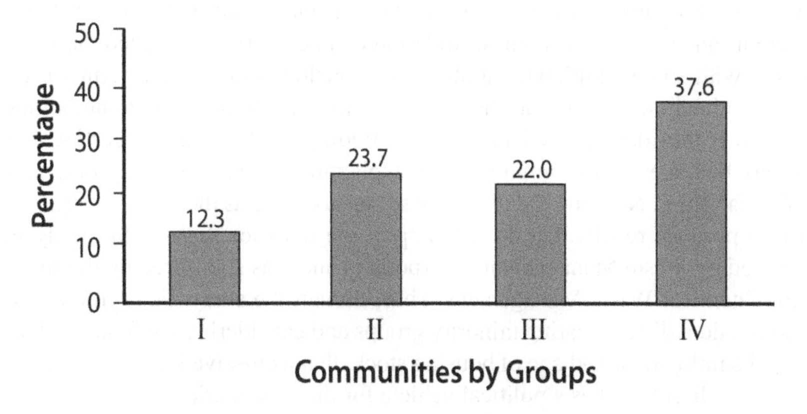

Figure 8.2 Percentage minority by exposure group in the U.S. 1986

Source: United Church of Christ (1987), reprinted by permission from the Commission for Racial Justice (United Church of Christ) Toxic Wastes and Race in the United States: A National Report on the Racial and Socio-Economic Characteristics of Communities with Hazardous Waste Sites. Reproduction of Original Graph, Minority* Percentage of the Population in U.S. Communities with Operating Commercial Hazardous Waste Facilities, Figure 1, Page 14. Copyright © 1987 United Church of Christ. All rights reserved.

other 'non-whites'). Indeed they found that zip codes with one or more hazardous waste sites had on average two times the percentage of these minority groups compared to zip codes without these environmental threats. One of the most influential state-level analyses in this period was by Mohai and Bryant (1992), which showed a dose-response relationship away from commercial hazardous waste facilities in the greater Detroit area with 48% African American within 1 mile of a facility, 39% in the 1- to 1.5-mile zone and only 18% in the farthest zone of 1.5+ miles. This latter study was considered an improvement on previous studies since it focused on the percentage minority at specific distances away from each facility to suggest a dose-response relationship, not simply whether a facility was somewhere in what often turned out to be very large rural counties. The problem of large spatial units is one of a number of critiques of these early studies.

Nevertheless, the power of these early studies cannot be understated as they were a key impetus to the 1994 Executive Order on environmental justice. These researchers had articulated what several had already known – the U.S. had a serious environmental injustice problem. Several spatial analyses followed along with some counterevidence (Anderton et al., 1994) and critiques of the spatial methodologies in particular (Bowen, 2002; Maantay, 2002). Among the main critiques is that the choice of spatial unit can cause results to flip-flop such that environmental injustice indicated in studies using large units like zip codes or counties may disappear when smaller divisions like census tracts are used as the unit of spatial analysis (Bowen et al., 1995; Maantay, 2002). This is a problem well known to geographers and spatial analysts as the modifiable areal unit problem (MAUP) when findings change depending on the spatial unit of analysis identified. Problems such as this bring into focus the notion that analytical choices are not necessarily immune to politics (see also Chapter 2). This is in fact one of the problems with Anderton et al.'s (1994) critique against the USGAO and UCC environmental justice studies. Anderton et al. essentially repeated the UCC study but with some additional variables and concluded that African Americans were not unjustly exposed, but Hispanics and low-dwelling-value communities were. Further they downplayed any injustice findings in the study suggesting that other variables were more important for explaining the patterns of people and polluting facilities. Not only did they choose census tracts as the unit of analysis – known to produce results that do not support the injustice thesis – the analysis was funded by Waste Management Incorporated and was sponsored by the Institute for Chemical Waste Management. Thus, though the study did highlight the need to consider differentiating minority groups and considering confounders like industrial employment and age of housing stock, the dismissive interpretation has been largely discredited as a political vehicle for these industries.

Other researchers are less concerned with confounders and more concerned with increasing the accuracy of the exposure measures. For example, within a large spatial unit like a county or zip code area minority populations and waste sites are not evenly distributed – a problem labelled the *unit-hazard coincidence problem* (Mohai and Saha, 2006). That is, the facility may be in one part of the spatial unit while the highest density of minority population is several kilometres away in a different part of the spatial unit. Nevertheless, when such spatial coincidence occurs within the spatial unit, it would be counted as a case of injustice using the 1987 UCC methodology (Figure 8.3). Of course, if there is a high percentage minority or low-income groups near the facility but located in the neighbouring spatial unit, then the people in that neighbouring unit would not be considered to be exposed using the hazard-unit coincidence method. When these effects are not random a study will report high false positives or false negatives. That is why the unit coincidence method is now considered useful mainly as a first cut at detecting patterns of environmental injustice that require more refined exposure analysis.

As mentioned, one solution to this problem has been to simply define populations outwards from the facilities themselves – something Mohai and Bryant (1992) used to show the distance decay of the percentage African American residents away from hazardous waste treatment, storage and disposal facilities (TSDFs) in Michigan. Geographic information systems make it relatively simple to create 'buffers' around facilities that can then be used to apportion out exposure to the spatial units as in Figure 8.3.

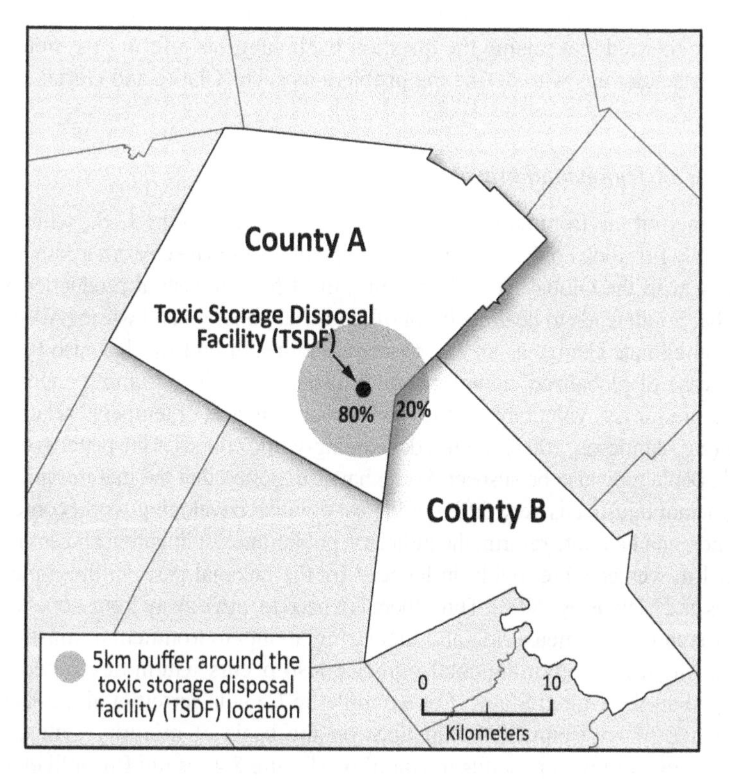

Figure 8.3 The unit-hazard coincidence problem

Source: Adapted from Mohai and Saha (2006).

The race or class debate

An early avenue of enquiry and source of political debate revolved around the idea that race may not be as important in patterns of environmental injustice as social class. Yet the UCC (1987, p. 13) study findings did not support this idea – one of the major conclusions of that study was that "mean household income and mean value of owner-occupied homes were not as significant as the mean minority percentage". Others have suggested that income and dwelling value as proxies for class are more significant than race as a predictor of toxic exposure. Environmental justice research has evolved much more recently in places like the UK with the emphasis of both research and policy on class rather than race or ethnicity (Mitchell and Dorling, 2003; Bulkeley and Walker, 2005; Walker et al., 2005). Similarly recent research in Canada falls somewhere between and includes studies which measure both social class and ethnicity in the same studies (Jerrett et al., 2001; Buzzelli and Jerrett, 2004; Agyeman et al., 2010). Mantaay (2001) points out a number of studies have shown a strong 'class' effect while others have shown strong minority population effects, leading her to conclude that the debate is unresolvable. Further she concurs with those who remind that minority status, income and wealth are so heavily intertwined that it is not particularly useful to devise analytical strategies that would suggest otherwise. These same writers are also concerned that raising the question itself suggests a form of systemic racism among academics who define the problem as such (Clarke and Gerlak, 1998).

Global environmental injustice

Evidence of environmental injustice is not confined to the U.S., where the idea arguably first took hold. It is almost trite to point out global-scale environmental justice issues, with the Global North benefitting most from industrial production while the Global South tends to bear the brunt of industrial pollution and waste (Adeola, 2000). Global climate change as an environment and health problem has also been framed as a case of globalized environmental injustice, as high-consuming countries in the north are far less vulnerable to the more serious impacts of temperature and sea-level rise (e.g., Jamieson, 2001). Within developing-world countries the patterns of pollution and people may also be suspect. Some have suggested that the manifestation of environmental injustice is very different in post-colonial developing-world contexts where phenomena like state reform, the judiciary, public-interest litigation and environmental social movements are highly influenced by the colonial pasts in these places (Williams and Mawdsley, 2006). Thus, there is a need to move away from a one-size-fits-all approach to both measuring and theorizing about environmental injustice.

As suggested, environmental injustice also occurs within developed countries other than the United States. For example, in Canada Jerrett et al. (2001) showed evidence of environmental injustices on the basis of minority status, dwelling value and employment status in Hamilton (Figure 8.4) as did Buzzelli et al. (2006) in Vancouver. In the Australian context Lloyd-Smith and Bell (2003) provide two case examples of communities with high minority populations struggling with toxic waste problems in their account of the "rise of environmental justice". In the

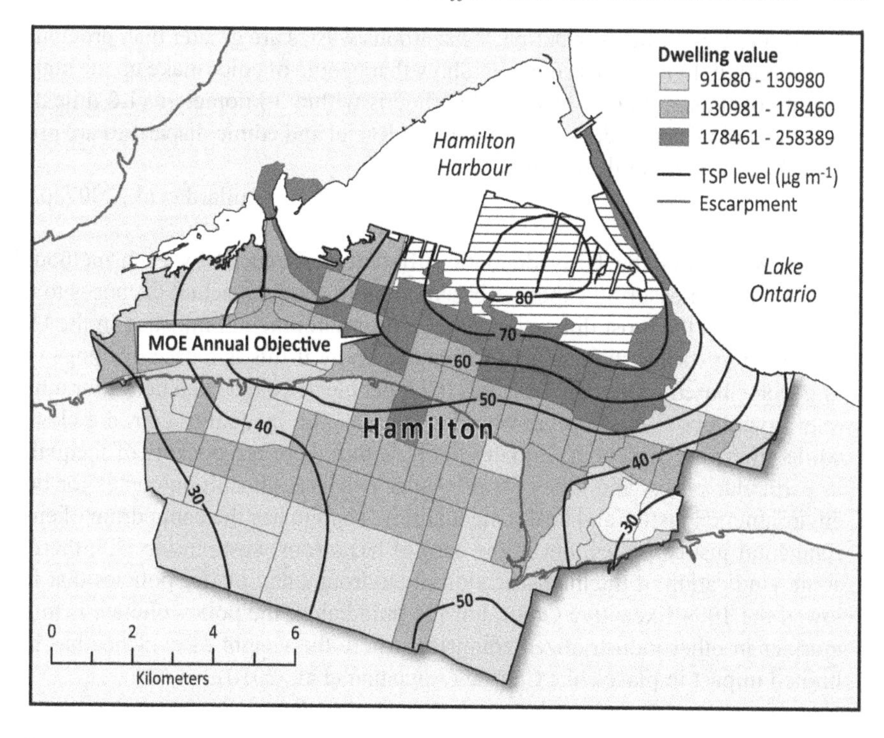

Figure 8.4 Environmental injustice in Hamilton, Canada

Source: Jerrett et al. (2001) open access.

UK Mitchell and Dorling (2003) use car ownership as a proxy for social class and find that those who own the most and largest vehicles are the least likely to live in high-traffic-pollution areas while those who do not own vehicles are most exposed. These studies help push environmental justice onto national policy agendas by articulating the conditions under which injustice occurs using a range of pollution exposures. In some cases there may be extensive patterns of environmental injustice (e.g., Mitchell and Dorling, 2003), yet there may also be individual cases in particular places that deserve in-depth attention (Lloyd-Smith and Bell, 2003).

Environmental injustice over time

The United Church of Christ in the U.S. was asked to revisit the issue of environmental justice by updating their methodology and addressing research problems outlined by spatial analysts (e.g., Bowen, 2002). Another key aim of the update was to determine if the situation is getting worse or better over time in the U.S. In their publication titled *Toxic Wastes and Race at Twenty 1987–2007* they concluded that in fact the situation had worsened:

> The application of these new methods, which better determine where people live in relation to where hazardous sites are located, reveals that racial

> disparities in the distribution of hazardous wastes are greater than previously reported. In fact, these methods show that people of color make up the majority of those living in host neighborhoods within 3 kilometers (1.8 miles) of the nation's hazardous waste facilities. Racial and ethnic disparities are prevalent throughout the country.
>
> (Bullard et al., 2007, p. x)

Figure 8.5 shows the fairly dramatic differences produced by each methodology whereby measuring exposure based on more rigorous actual distances to the facility yields the most dramatic evidence of environmental injustice in the U.S. compared to earlier studies. It is difficult to know if this is a case of things getting worse based on certain measures (e.g., people of colour) and better for others (e.g., social class) in the same way that some health inequality gaps are closing while others are widening. Equally likely is that there are pockets of inequality in particular places and times for particular exposures/health outcomes for both environmental justice and health inequalities. Therein lies the conundrum of environmental justice policy, but in the case of hazardous waste in the U.S. there is some vindication of the intensification of environmental justice policies that followed the 1994 Executive Order. For the time being, the policy climate is much murkier in other industrialized countries, where the weight of evidence has had limited impact in places like Canada (Agyeman et al., 2010).

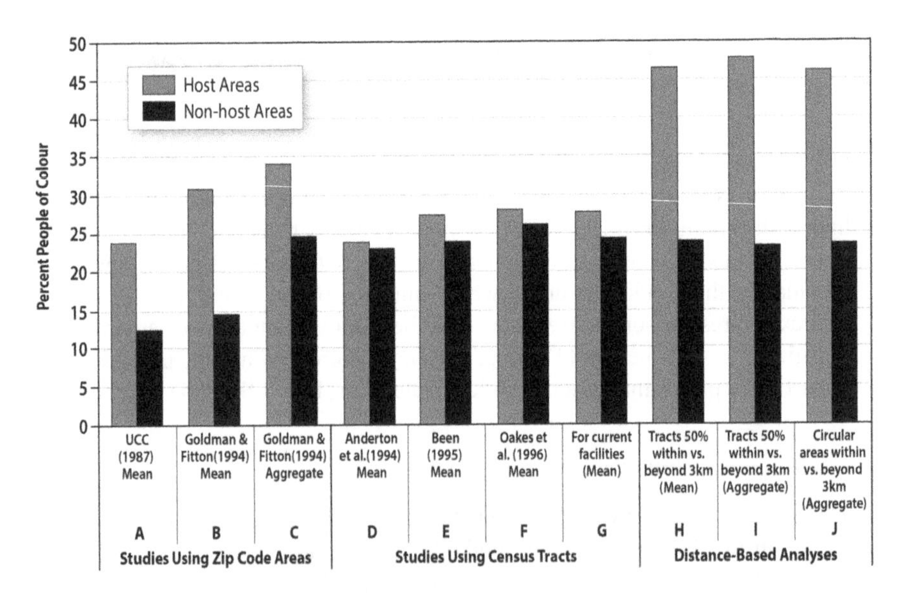

Figure 8.5 Unit-hazard coincidence vs. distance

Source: Bullard et al. (2007), reprinted by permission from Toxic Wastes and Race at Twenty 1987–2007: A Report Prepared for the United Church of Christ Justice & Witness Ministries, Reproduction of Original Graph, Comparing Results of Past Studies Using Unit-Hazard Coincidence Method with Results Using Distance-Based Methods, Figure 3.3, Page 44.

Procedural environmental equity and justice

Accepting that instances of environmental injustice do indeed exist in one form or another leaves the more challenging issues of determining the social, cultural and above all institutional processes that cause and sustain these spatial patterns and using that information to address the problem. Yet the study of procedural justice as an area of enquiry is relatively less well developed, or at least diffuse in the academic literature – found in several corners of the social sciences relating to a range of issues including social justice, civil rights, racism, cultural studies, governance, civil society and First Nations studies. These sorts of academic research and writing may not specifically use the term 'environmental justice' or 'environmental equity', but they may nevertheless concern the social and institutional process that lead to disproportionate pollution exposure. The linkages between those disparate literatures and the environmental justice literature continue to grow (Agyeman et al., 2010). Given the potential breadth of this area, three areas of enquiry are elaborated here to provide a sense of the issues involved – facility siting, market mechanisms and mitigation monitoring and emergency response. A main theme is that deep, socially embedded structures of power are at the root of most explanations.

Facility siting fairness

A simple way to explain how patterns of injustice emerge is that the siting process is flawed and unfair. That is, facility owners look for the path of least resistance, which includes communities that likely will not or cannot adequately mount a successful campaign against the facility or project. Even when low-income and minority populations are highly involved and supportive of each other they may not have the specific tools required to stop a project. One of the most challenging problems is the issue of *public participation* in the siting process, particularly the *environmental impact assessment process*. Arnstein (1969) points out that the problem with public participation is power in decision making – that the potential participants often do not feel they are allowed to play a meaningful role. Residents may feel they are merely being informed about decisions that have already being made with no opportunity to shape the form of the development (Smith and McDonough, 2001). When a community is not allowed to make the case to prevent a development on the grounds of environmental injustice, they may feel they are simply being used – a process Arnstein labels *tokenism*. Though Dodds and Hopwood (2006) provide a rare counter-example of successful working-class opposition to a replacement waste incinerator in Newcastle, England, they point out that it is more often the case that such a community simply may not have the resources available to make the case that a proposed new development represents an environmental injustice.

In the most charitable view of environmental assessment processes, the structures for participation are in place and fair to allow all local groups to bring forward reasonable arguments for not locating a facility locally, but locals simply cannot find the time or resources to participate. Yet others have suggested that the existing official arrangements for participation are flawed and biased and disproportionately favour those who are already highly connected to structures of power at several scales. Innes and Booher (2004) argue that in the U.S. context participatory processes often have a net damaging effect due to the loss of trust, which resonates with Deacon and Baxter's (2013) Nova Scotia, Canada, case-study finding that the process is often highly technical, with few opportunities to even say no to a facility. If community members feel their time is being wasted through tokenism, that their input is not being heard, anger and frustration are the results, leading to an overall loss of trust in the system (O'Faircheallaigh, 2010).

There is a more insidious cumulative process potentially at work leading to injustices that may not be readily addressed by project-by-project environmental assessment (Corburn, 2002; Fox, 2002). The Camden case led to the New Jersey courts requiring that opposition groups prove that facility owners *intentionally* picked low-income, minority neighbourhoods for their facilities because these areas are already marginalized. This is a difficult standard to meet. There need not be intentionality per se if long and drawn-out siting processes devolve into a retooling to prioritize avoidance of opposition over other criteria for site appropriateness. That is, if less vulnerable, wealthier neighbourhoods repeatedly oppose facilities successfully, patterns of injustice may also emerge without demonstrable or specific intent on the part of developers. In such a scenario, devising policy to combat this process is particularly challenging but likely has to come from direct government intervention.

Market mechanisms and the 'which came first' debate

What if minority, low-income and other vulnerable groups moved into homes near a facility after it became operational? Does it even matter from a justice viewpoint? Some have suggested that this would mean that the predominant process is a move-in effect rather than systemic siting biases and that the mechanism or 'cause' is essentially market forces (Anderton et al., 1994). For example, Mitchell et al. (1999) used geographic information systems to analyze the coincidence of race, social class and large facilities that emit toxic chemicals using the Toxic Release Inventory (TRI) in South Carolina. They found that over time the percentage of minority and low-income residents tended to increase around the facilities, lending support to the move-in thesis. Others have provided similar but longitudinal evidence and have suggested that it is not unfair siting processes that lead to injustices but that facilities lower property values in the surrounding neighbourhoods (Been, 1994; Been and Gupta, 1997). Justice advocates have suggested instead that all of this research is a distraction from more important issues.

Some justice advocates point out, for example, that housing 'market forces' are not immune to systemic structural biases that limit choices for vulnerable groups. The controversial practice of redlining in the real estate market (refusing mortgage lending in some neighbourhoods but not others) is one well-known example of structural injustices that can emerge in the marketplace (Massey and Denton, 1993).

Blaming the patterns on the market or worse, the vulnerable groups themselves that supposedly have the free choice to live wherever they please masks more deeply entrenched structural problems. Further, Pulido (1996) has pointed out that the projects against both environmental racism and environmental justice more broadly have been distracted from giving due attention to the economic, historic, political and social structures that reproduce racial and other tightly intertwined forms of (environmental) injustice. She argues that academics have been distracted too much by such 'chicken or egg' debates including teasing out how much is social class and how much is race (see also Maantay, 2002). On the other hand Marshall (2010) suggests working outside the typical legal and environmental assessment frameworks. She reviews the strategies grassroots environmental justice activists use at the local level. She highlights how lawyers within the environmental justice movement tend to recommend legal tactics mainly to build support for the local political movement, while they caution against the use of legal procedures to invoke environmental change directly within existing systems. These lawyers tend to suggest that media campaigns and non-violent civil protest are more effective at provoking meaningful change. Thus, both Pulido and Marshall caution that giving the local and wider environmental justice movements over to existing legal and scientific 'procedures' is problematic. These authors highlight that environmental justice is tied to broader issues of social justice and the political structures that have long sustained marginalization of low-income and racial/ethnic minority groups.

Mitigation, monitoring and emergency response

Procedural environmental justice may also refer to the processes of mitigation and monitoring of existing facilities. *Mitigation* refers to measures meant to reduce the negative impacts of a facility and may include such measures as berms, pollution control measures, special truck routes or even subsidies for air conditioners so windows can be kept closed. Mitigation may also refer to cleanup in cases where accidental leaks happen. *Monitoring* refers to ongoing measurements to ensure that negative impacts are detected then minimized. This could involve a blood or hair monitoring program to ensure that residents and workers are not exposed to levels of toxins that exceed guideline amounts (Shepard, 2002).

Justice in emergency response refers to equal treatment in the face of environmental disasters that may or may not be attributable to any particular point source. The case of Hurricane Katrina (2005) is a recent North American example in which emergency response to the disaster and flooding largely failed low-income, African American and other communities of colour in numerous ways. Not only were several people unaccounted for weeks into the disaster, but emotional support, job loss, returning residents to their neighbourhood and overall recovery have been unjust along both racial and class lines. Low-income African American homeowners have been identified as being particularly hardest hit (Elliott and Pais, 2006; Bullard, 2009; Bullard and Wright, 2010). Though such studies point out the positive impact of deeply engrained family and community support, they simultaneously highlight how fragile such networks can be in the face of disaster.

Dilution of the meaning of environmental justice

One of the challenges for environmental justice as both a political movement and an area of academic enquiry is that the term is used so frequently now that its meaning and therefore its power may get watered down. In the developed world in particular cases of opposition to potentially toxic facilities are quite familiar. The challenge lies in the fact that often the community opposed to the facility would not be defined as vulnerable and may have a relatively small percentage of minority and low-income residents. That is, environmental justice may be used as a rhetorical tool to fight these facilities. Politicians in particular have identified this as a problem of 'not in my back yard' (NIMBY), a pejorative concept suggesting that those opposed to hazardous facilities in their local community are otherwise supportive of the technology. Being labelled NIMBY represents an accusation of the failure to do one's part to manage the environmental bads in society. NIMBY has thus been used rhetorically to categorically discredit any opposition to unwanted facilities (McGurty, 1997), including that from socially disadvantaged groups. The research evidence suggests though that opposition labelled as NIMBY is often associated with parallel claims of unfair siting processes that do not adequately engage local communities (Wolsink, 2006). Thus, there is a tension between the environmental justice explanation and the NIMBY explanation for resistance to facilities. Lake (1996) tends to associate NIMBY with wealthier communities that, by avoiding potentially noxious facilities, put less powerful, marginalized communities at risk. The issue of environmental justice has become more complex as communities of all sorts are claiming to be victims of pollution injustices, including wealthy non-minority. Arguments against such facilities may be on the grounds of distributive (outcome) equity if the community has already hosted a facility in the past rather than fairness of process (procedural; e.g., if alternatives to minority and low-income community sites were not considered). Such issues move away from environmental justice as originally conceived and may ironically lead to sidestepping the plight of groups that are already marginalized. The growing use of the term 'environmental justice' across a wide array of contexts greatly complicates how policy and policy actors might deal with environmental justice moving forward.

Environmental justice and policy

What is to be done to properly deal with procedural and outcome environmental injustice? Many jurisdictions are only beginning to think about how to incorporate environmental justice into laws, guidelines and decision making. The United States is taking a lead role in this regard, with not only several policies on environmental justice but also one of the highest concentrations of non-governmental activist networks on the topic. Not only does the U.S. have federal legislation meant to combat environmental injustices, several states have enacted their own legislation to align with these federal laws. More recently, the European Union in 1998 similarly adopted the Access to Information, Public Participation in Decision-making and Access to Justice in Environmental Matters or Aarhus Convention, which outlines principles of procedural justice in the EU context to ensure citizens have access to publicly available data and decision making. Contrast this with Canada, which has no mention of environmental justice in its key pieces of environmental protection and environmental assessment legislation, with most provinces similarly lacking specific recognition of environmental justice in their environmental laws (Agyeman et al., 2010).

A cornerstone of the EPA's efforts on behalf of the U.S. federal government is to follow the principle of 'right to know'. Before a community can make a claim of environmental injustice they must first know what they are being exposed to and what negative outcomes are being experienced at a neighbourhood/community level (Lambert et al., 2003). It is for this reason that pollution release information is now routinely reported and made publicly available. For example, the federally mandated Toxic Release Inventory (TRI) keeps track of the substance releases of all major polluting facilities in the U.S., while the National Pollutant Release Inventory performs the same function in Canada. Likewise efforts have been made to make health (mostly disease) and socio-demographic (e.g., census) data publicly available at various spatial scales. In a study of the first decade after the implementation of the TRI in the U.S. Grant (1997) finds that those states with funding to support and extend the scope of right-to-know legislation or have right-to-sue legislation have lower levels of pollution than states that do not.

Though data available through right-to-know have been used effectively to show that environmental injustices persist (e.g., Bowen et al., 1995), data alone are not enough. What are also required are tools for bringing these data together in formats that are easily accessed and understood. For example, the Environmental Protection Agency (EPA) in the U.S. provides a geographic information system (GIS) tool called EJSCREEN (EPA, 2016) which is an example of how data sets may be brought together to understand if there is cause to be concerned about pollution leading to elevated outcomes like elevated cancer risk. Data sets in EJSCREEN include information like hazardous waste and Superfund sites as proxies for toxic exposure. Though such tools are highly sophisticated, allowing the user to change the scale with the roll of a mouse wheel or click of a button, it still requires access to the computing resources needed to run these web-based tools. EJSCREEN and similar tools may be used as inputs for assessing a community's environmental injustice claim, which can in turn be used to evaluate things like Superfund priority or qualification for a range of environmental justice grants if available. Such grants can be used to mitigate cumulative effects like contaminated brownfields or contribute towards empowering communities to be more thoroughly involved in local environmental assessments and other forms of environmental decision making.

A more controversial mechanism for dealing with environmental inequities is to compensate for harms (Boerner and Lambert, 1995). The facility owner either pays local residents directly on a household-by-household basis or collectively using an in-kind arrangement in terms of providing funds for community projects (e.g., recreational facilities, community centre). For example, a deep geologic repository for radioactive waste in Canada is accompanied by $35 million in financial benefits (Castellan and Barker, 2006). Such an approach is controversial for two key reasons. First, such compensation does not directly address the problem of exposure to contaminants; it merely offsets the negative costs by providing benefits (Mank, 1995). Thus, the residents will remain disproportionately exposed and potentially at higher risk for negative health outcomes from toxic chemicals. Second, offering compensation may actually exacerbate the negative psychosocial impacts of the facility if the offer is perceived by residents to be a bribe. In many cases what the residents want most is to participate in decision making and know that facility design, mitigation and monitoring will ensure the safety of the facility is acceptably high. When pay-outs are offered too early residents may see this as a signal of unwillingness on the part of facility owners to engage in these more important activities (Pellow et al., 2001). One of the justice challenges for developers then is to ensure that compensation is not a substitute for minimizing facility risks.

Conclusion

The inequalities in health and environmental justice literatures intersect and need to be made more explicit. That is, the inequalities in health literature hint at underlying environmental differences as a cause of the social gradient in health; the environmental justice literature similarly suggests linkages between disproportionate pollution exposure (or therapeutic environments), vulnerable social groups and disease outcomes. Moving forward we will likely see tighter linkages between these literatures (see Masuda et al., 2010; Wakefield and Baxter, 2010). Yet *disease* does not need to be the ultimate consideration for motivating policy action. For example, exposure to higher levels of total suspended particulates may result in increased cardiovascular or pulmonary disease burden for poorer and minority populations; but the disproportionate exposure itself should be ample reason to promote remediation actions since pollution represents an added burden to these groups. Higher exposure to polluted environments or lack of access to positive physical environments may have their most profound effects on overall well-being whether or not that is connected to specific somatic effects. This underscores the fact that epidemiology, toxicology and geographical spatial analyses – outcome justice – only tell part of the story of environmental justice. Conversely, focussing simply on fair process without due attention to spatial and social distribution of environmental pollution and negative health outcomes is problematic. While the U.S. has arguably bolstered its procedural environmental justice apparatus over the last couple of decades, some patterns of injustice have worsened in the same time period. This reminds us that there are political and social structures that need to be better understood to address procedural justice issues. Spatial patterns are best addressed by quantitative and spatial social science research, but procedural justice may be more appropriately addressed by critical and qualitative approaches to social scientific environment and health enquiry. Many of these issues are taken up in

Section III and particularly Chapter 9, where the social and political process at work in pollution exposure, impacts, concern and action are further explored.

Note

1 Superfund sites are large contaminated sites in need of extensive cleanup to safeguard human health. The fund is enabled by the Comprehensive Environmental Response, Compensation, and Liability Act of 1980 (CERCLA) and is enacted largely by the U.S. Environmental Protection Agency (EPA). Though sizable, the fund is limited, and there is much political debate over the identification and prioritization of sites for cleanup.

References

Adeola, F.O. 2000. Cross-national environmental injustice and human rights issues a review of evidence in the developing world. *American Behavioral Scientist*, 43(4), 686–706.

Agyeman, J., Bullard, R.D., & Evans, B. 2003. *Just Sustainabilities: Development in an Unequal World.* London: Earthscan.

Agyeman, J., Cole, P., Haluza-DeLay, R., & O'Riley, P. (Eds.). 2010. *Speaking for Ourselves: Environmental Justice in Canada.* Vancouver, BC: University of British Columbia Press.

Anderton, D.L., Anderson, A.B., Oakes, J.M., & Fraser, M.R. 1994. Environmental equity: The demographics of dumping. *Demography*, 31(2), 229–248.

Arnstein, S.R. 1969. A ladder of citizen participation. *Journal of the American Institute of Planners*, 35(4), 216–224.

Atari, D.O., Luginaah, I., & Baxter, J. 2011. 'This is the mess that we are living in': Residents everyday life experiences of living in a stigmatized community. *GeoJournal*, 76(5), 483–500.

Atari, D.O., Luginaah, I.N., Gorey, K., Xu, X., & Fung, K. 2013. Associations between self-reported odour annoyance and volatile organic compounds in 'Chemical Valley', Sarnia, Ontario. *Environmental Monitoring and Assessment*, 185(6), 4537–4549.

Been, V. 1994. Locally undesirable land uses in minority neighborhoods: Disproportionate siting or market dynamics? *Yale Law Journal*, 103(6), 1383–1422.

Been, V., & Gupta, F. 1997. Coming to the nuisance or going to the barrios-A longitudinal analysis of environmental justice claims. *Ecology Law Quarterly*, 24, 1–56.

Bleich, S.N., Jarlenski, M.P., Bell, C.N., & LaVeist, T.A. 2012. Health inequalities: Trends, progress, and policy. *Annual Review of Public Health*, 33, 7–40.

Boerner, C., & Lambert, T. 1995. Environmental injustice. *Public Interest*, 118, 61–82.

Bowen, W. 2002. An analytical review of environmental justice research: What do we really know? *Environmental Management*, 29(1), 3–15.

Bowen, W.M., Salling, M.J., Haynes, K.E., & Cyran, E.J. 1995. Toward environmental justice: Spatial equity in Ohio and Cleveland. *Annals of the Association of American Geographers*, 85(4), 641–663.

Brulle, R.J., & Pellow, D.N. 2006. Environmental justice: Human health and environmental inequalities. *Annual Review Public Health*, 27, 103–124.

Buckingham-Hatfield, S., Reeves, D., & Batchelor, A. 2005. Wasting women: The environmental justice of including women in municipal waste management. *Local Environment*, 10(4), 427–444.

Bulkeley, H., & Walker, G. 2005. Environmental justice: A new agenda for the UK. *Local Environment*, 10(4), 329–332.

Bullard, R.D. 2009. *Race, Place, and Environmental Justice after Hurricane Katrina: Struggles to Reclaim, Rebuild, and Revitalize New Orleans and the Gulf Coast*. Boulder, CO: Westview Press.

Bullard, R.D., Mohai, P., Saha, R., & Wright, B. 2007. *Toxic Wastes and Race at Twenty 1987–2007: Grassroots Struggles to Dismantle Environmental Racism in the United States*. Cleveland, OH: United Church of Christ Justice and Witness Ministry.

Bullard, R.D., & Wright, B. 2010. *Race, Place, and Environmental Justice after Hurricane Katrina: Struggles to Reclaim, Rebuild, and Revitalize New Orleans and the Gulf Coast (Large Print 16 pt)*. ReadHowYouWant.com.

Buzzelli, M., & Jerrett, M. 2004. Racial gradients of ambient air pollution exposure in Hamilton, Canada. *Environment and Planning A*, 36(10), 1855–1876.

Buzzelli, M., Su, J., Le, N., & Bache, T. 2006. Health hazards and socio-economic status: A neighbourhood cohort approach, Vancouver, 1976–2001. *The Canadian Geographer/ Le Géographe Canadien*, 50(3), 376–391.

Castellan, A.G., & Barker, D.E. 2006. 'The OPG/Kincardine Hosting Agreement for a Deep Geologic Repository for OPG's Low-and Intermediate-Level Waste'.

Centers for Disease Control (CDC). 2015. *Black or African American Populations*. Available at: http://www.cdc.gov/minorityhealth/populations/REMP/black.html (Accessed 7 January 2015).

Clarke, J., & Gerlak, A. 1998. 'Environmental Racism in Southern Arizona.' in Camacho, D.E. (ed.). *Environmental Injustices, Political Struggles: Race, Class, and the Environment*. Durham, NC: Duke University Press. pp 82–100.

Corburn, J. 2002. Environmental justice, local knowledge, and risk: The discourse of a community-based cumulative exposure assessment. *Environmental Management*, 29(4), 451–466.

Cutter, S.L. 1995. Race, class and environmental justice. *Progress in Human Geography*, 19, 111–122.

Deacon, L., & Baxter, J. 2013. No opportunity to say no: A case study of procedural environmental injustice in Canada. *Journal of Environmental Planning and Management*, 56(5), 607–623.

Dobson, A. 1998. *Justice and the Environment: Conceptions of Environmental Sustainability and Dimensions of Social Justice*. London: Oxford University Press.

Dodds, L., & Hopwood, B. 2006. BAN waste, environmental justice and citizen participation in policy setting. *Local Environment*, 11(03), 269–286.

Dunion, K., & Scandrett, E. 2003. 'The Campaign for Environmental Justice in Scotland as a Response to Poverty in a Northern Nation.' in Bullard, R.D., Agyeman, J., & Evans, R. (eds.). *Just Sustainabilities: Development in an Unequal World*. London: Earthscan. pp 311–322.

Elliott, J.R., & Pais, J. 2006. Race, class, and Hurricane Katrina: Social differences in human responses to disaster. *Social Science Research*, 35(2), 295–321.

Environmental Protection Agency (EPA). 2016. *EJSCREEN*. Available at: http://www.epa.gov/ejscreen (Accessed 19 January 2016).

Foster, S. 2004. The challenge of environmental justice. *Rutgers University Journal of Law and Public Policy*, 1(1), 6–19.

Fox, M.A. 2002. Evaluating cumulative risk assessment for environmental justice: A community case study. *Environmental Health Perspectives*, 110(Suppl 2), 203–209.

Galli, A., Kitzes, J., Niccolucci, V., Wackernagel, M., Wada, Y., & Marchettini, N. 2012. Assessing the global environmental consequences of economic growth through the ecological footprint: A focus on China and India. *Ecological Indicators*, 17, 99–107.

Gee, G. 2002. A multilevel analysis of the relationship between institutional and individual racial discrimination and health status. *American Journal of Public Health*, 92(4), 615–623.

Gesler, W.M. 1992. Therapeutic landscapes: Medical issues in light of the new cultural geography. *Social Science & Medicine*, 34(7), 735–746.

Gottfredson, L. 2004. Intelligence: Is it the epidemiologists' elusive "fundamental cause" of social class inequalities in health? *Journal of Personality and Social Psychology*, 86(1), 174–199.

Grant, D.S. 1997. Allowing citizen participation in environmental regulation: An empirical analysis of the effects of right-to-sue and right-to-know provisions on industry's toxic emissions. *Social Science Quarterly*, 78(4), 859–873.

Hoover, E., Cook, K., Plain, R., Sanchez, K., Waghiyi, V., Miller, P., . . . & Carpenter, D.O. 2012. Indigenous peoples of North America: Environmental exposures and reproductive justice. *Environmental Health Perspectives*, 120(2), 1645–1649.

Huurre, T., Aro, H., & Rahkonen, O. 2003. Well-being and health behaviour by parental socioeconomic status. *Social Psychiatry and Psychiatric Epidemiology*, 38(5), 249–255.

Innes, J. E., & Booher, D. E. 2004. Reframing public participation: Strategies for the 21st century. *Planning Theory & Practice*, 5(4), 419–436.

Jamieson, D. 2001. Climate change and global environmental justice. In Miller, C.A. and Edwards, P.N. (eds.). *Changing the Atmosphere: Expert Knowledge and Global Environmental Governance*. Cambridge: MIT Press, pp 287–307.

Jerrett, M., Burnett, R.T., Kanaroglou, P., Eyles, J., Finkelstein, N., Giovis, C., & Brook, J.R. 2001. A GIS-environmental justice analysis of particulate air pollution in Hamilton, Canada. *Environment and Planning A*, 33(6), 955–974.

Kaplan, G., Pamuk, E., Lynch, J., Cohen, R., & Balfour, J. Apr 20 1996. Inequality in income and mortality in the United States: Analysis of mortality and potential pathways. *British Medical Journal*, 312, 999–1003.

Krieger, N. 2012. Methods for the scientific study of discrimination and health: An ecosocial approach. *American Journal of Public Health*, 102(5), 936–944.

Krieger, N., Williams, D., & Moss, N. 1997. Measuring social class in US public health research: Concepts, methodologies, and guidelines. *Annual Review of Public Health*, 18, 341–378.

Laaksonen, M., Rahkonen, O., Karvonen, S., & Lahelma, E. 2005. Socioeconomic status and smoking: Analysing inequalities with multiple indicators. *European Journal of Public Health*, 15(3), 262–269.

Lake, R.W. 1996. Volunteers, NIMBYs, and environmental justice: Dilemmas of democratic practice. *Antipode*, 28(2), 160–174.

Lambert, T.W., Soskolne, C.L., Bergum, V., Howell, J., & Dossetor, J.B. 2003. Ethical perspectives for public and environmental health: Fostering autonomy and the right to know. *Environmental Health Perspectives*, 111(2), 133–137.

Lloyd-Smith, M.E., & Bell, L. 2003. Toxic disputes and the rise of environmental justice in Australia. *International Journal of Occupational and Environmental Health*, 9(1), 14–23.

Maantay, J. 2001. Zoning, equity, and public health. *American Journal of Public Health*, 91(7), 1033.

Maantay, J. 2002. Mapping environmental injustices: Pitfalls and potential of geographic information systems in assessing environmental health and equity. *Environmental Health Perspectives*, 110(Suppl 2), 161–171.

Macintyre, S., Ellaway, A., & Cummins, S. 2002. Place effects on health: How can we conceptualise, operationalise and measure them? *Social Science & Medicine*, 55(1), 125–139.

Mackenbach, J.P. 2012. The persistence of health inequalities in modern welfare states: The explanation of a paradox. *Social Science & Medicine*, 75(4), 761–769.

Mank, B.C. 1995. Environmental justice and discriminatory siting: Risk-based representation and equitable compensation. *Ohio State Law Journal*, 56, 329–425.

Marmor, T.R., Barer, T.L., & Evans, R.G. (Eds.). 1994. *Why Are Some People Healthy and Others Not*. New York: Aldine de Gruyter.

Marmot, M. 2001. Inequalities in health. *New England Journal of Medicine*, 345(2), 134–135.

Marmot, M. 2005. Social determinants of health inequalities. *The Lancet*, 365(9464), 1099–1104.

Marmot, M., & Brunner, E. 2005. Cohort profile: The Whitehall II study. *International Journal of Epidemiology*, 34(2), 251–256.

Marmot, M., Stansfeld, S., Patel, C., North, F., Head, J., White, I., . . . & Davey Smith, G. 1991. Health inequalities among British civil servants: The Whitehall II study. *The Lancet*, 337(8754), 1387–1393.

Massey, D.S., & Denton, N.A. 1993. *American Apartheid: Segregation and the Making of the Underclass*. Cambridge, MA: Harvard University Press.

Masuda, J.R., Poland, B., & Baxter, J. 2010. Reaching for environmental health justice: Canadian experiences for a comprehensive research, policy and advocacy agenda in health promotion. *Health Promotion International*, 25(4), 453–463. doi: 10.1093/heapro/daq041.

McGurty, E.M. 1997. From NIMBY to civil rights: The origins of the environmental justice movement. *Environmental History*, 2(3), 301–323.

Mitchell, G., & Dorling, D. 2003. An environmental justice analysis of British air quality. *Environment and Planning A*, 35(5), 909–929.

Mitchell, J.T., Thomas, D.S., & Cutter, S.L. 1999. Dumping in Dixie revisited: The evolution of environmental injustices in South Carolina. *Social Science Quarterly*, 80(2), 229–243.

Mitchell, R., & Popham, F. 2008. Effect of exposure to natural environment on health inequalities: An observational population study. *The Lancet*, 372, 1655–1660.

Mohai, P., & Bryant, B. 1992. Environmental injustice: Weighing race and class as factors in the distribution of environmental hazards. *University of Colorado Law Review*, 63(4), 921–932.

Mohai, P., & Saha, R. 2006. Reassessing racial and socioeconomic disparities in environmental justice research. *Demography*, 43(2), 383–399.

Navarro, V. 1990. Race or class versus race and class: Mortality differentials in the United States. *The Lancet*, 336(8725), 1238–1240.

Nazroo, J.Y. 2003. The structuring of ethnic inequalities in health: Economic position, racial discrimination, and racism. *American Journal of Public Health*, 93(2), 277–284.

O'Faircheallaigh, C. 2010. Public participation and environmental impact assessment: Purposes, implications, and lessons for public policy making. *Environmental Impact Assessment Review*, 30(1), 19–27.

Pellow, D.N., Weinberg, A., & Schnaiberg, A. 2001. The environmental justice movement: Equitable allocation of the costs and benefits of environmental management outcomes. *Social Justice Research*, 14(4), 423–439.

Pickett, K.E., & Pearl, M. 2001. Multilevel analyses of neighbourhood socioeconomic context and health outcomes: A critical review. *Journal of Epidemiology and Community Health*, 55(2), 111–122.

Pulido, L. 1996. A critical review of the methodology of environmental racism research. *Antipode*, 28(2), 142–159.

Raphael, D. 2000. Health inequalities in Canada: Current discourses and implications for public health action. *Critical Public Health*, 10(2), 193–216.

Rogers, R. 1992. Living and dying in the USA: Sociodemographic determinants of death among blacks and whites. *Demography*, 29, 287–303.

Ross, N.A., Wolfson, M.C., Dunn, J.R., Berthelot, J.M., Kaplan, G.A., & Lynch, J.W. 2000. Relation between income inequality and mortality in Canada and in the United States: Cross sectional assessment using census data and vital statistics. *British Medical Journal*, 320(7239), 898–902.

Sampson, R.J., Morenoff, J.D., & Gannon-Rowley, T. 2002. Assessing "neighborhood effects": Social processes and new directions in research. *Annual Review of Sociology*, 28, 443–478.

Scholsberg, D. 2004. Reconceiving environmental justice: Global movements and political theories. *Environmental Politics*, 13(3), 517–540.

Shepard, P.M., Northridge, M.E., Prakash, S., & Stover, G. 2002. Advancing environmental justice through community-based participatory research. *Environmental Health Perspectives*, 110(Suppl 2), 139–140.

Shippee, T.P., Rinaldo, L., & Ferraro, K.F. 2012. Mortality risk among Black and White working women: The role of perceived work trajectories. *Journal of Aging and Health*, 24(1), 141–167.

Siegrist, J., & Marmot, M. 2004. Health inequalities and the psychosocial environment – Two scientific challenges. *Social Science and Medicine*, 8, 1463–1473.

Silvern, S.E. 1999. Scales of justice: Law, American Indian treaty rights and the political construction of scale. *Political Geography*, 18(6), 639–668.

Sister, C., Wolch, J., and & Wilson, J., 2010. Got green? Addressing environmental justice in park provision. *GeoJournal*, 75(3), 229–248.

Smith, P. D., & McDonough, M. H. 2001. Beyond public participation: Fairness in natural resource decision making. *Society & Natural Resources*, 14(3), 239–249.

Subramanian, S., & Kawachi, I. 2004. Income inequality and health: What have we learned so far? *Epidemiologic Reviews*, 26, 78–91.

Swyngedouw, E., & Heynen, N.C. 2003. Urban political ecology, justice and the politics of scale. *Antipode*, 35(5), 898–918.

Taibbi, M. 2011. *Apocalypse, New Jersey: A Dispatch from America's Most Desperate Town.* Available at: http://www.rollingstone.com/culture/news/apocalypse-new-jersey-a-dispatch-from-americas-most-desperate-town-20131211 (Accessed 23 August 2014).

Taylor, D.E. 2000. The rise of the environmental justice paradigm: Injustice framing and the social construction of environmental discourses. *American Behavioral Scientist*, 43(4), 508–580.

United Church of Christ. Commission for Racial Justice. 1987. *Toxic Wastes and Race in the United States: A National Report on the Racial and Socio-Economic Characteristics of Communities with Hazardous Waste Sites*. Public Data Access. Available at: http://www.nrc.gov/docs/ML1310/ML13109A339.pdf (Accessed May 11, 2016).

U.S. General Accounting Office. 1983. *Siting of Hazardous Waste Landfills and Their Correlation with Racial and Economic Status of Surrounding Communities*. Washington, DC: Government Printing Office.

U.S. Government. 2017. Influenza.

van Dijk, T.K., Dijkshoorn, H., van Dijk, A., Cremer, S., & Agyemang, C. 2013. Multidimensional health locus of control and depressive symptoms in the multi-ethnic population of the Netherlands. *Social Psychiatry and Psychiatric Epidemiology*, 48(12), 1931–1939.

van Doorslaer, E., Wagstaff, A., Bleichrodt, H., Calonge, S., Gerdtham, U., Gerfin, M., . . . & Winkelhake, O. 1997. Income-related inequalities in health: Some international comparisons. *Journal of Health Economics*, 16, 93–112.

Venkatapuram, S. 2012. *Health Justice: An Argument from the Capabilities Approach*. Cambridge: Polity Press.

Wackernagel, M., & Rees, W. 1996. *Our Ecological Footprint: Reducing Human Impact on the Earth*. Gabriola Island, BC: New Society Publishers.

Wagstaff, A., Paci, P., & Van Doorslaer, E. 1991. On the measurement of inequalities in health. *Social Science & Medicine*, 33(5), 545–557.

Wakefield, S.E., & Baxter, J. 2010. Linking health inequality and environmental justice: Articulating a precautionary framework for research and action. *Environmental Justice*, 3(3), 95–102.

Walker, G. 2009. Beyond distribution and proximity: Exploring the multiple spatialities of environmental justice. *Antipode*, 41(4), 614–636.

Walker, G., Mitchell, G., Fairburn, J., & Smith, G. 2005. Industrial pollution and social deprivation: Evidence and complexity in evaluating and responding to environmental inequality. *Local Environment*, 10(4), 361–377.

Weestra, L. 2012. *Environmental Justice and the Rights of Indigenous People's: International and Domestic Legal Perspectives*. London: Earthscan.

Wiebe, S.M. 2012. 'Bodies on the Line: The In/Security of Everyday Life in Aamjiwnaang.' in Schnurr, M., & Swatuk, L. (eds.). *Natural Resources and Social Conflict: Towards Critical Environmental Security*. Basingstoke, UK: Palgrave MacMillan. pp 215–236.

Wilkinson, R. Feb 22 1997. Health inequalities: Relative or absolute material standards? *British Medical Journal*, 314, 591–595.

Williams, A. (Ed.). 2007. *Therapeutic Landscapes*. Burlington, VT: Ashgate.

Williams, D.R., Lavizzo-Mourey, R., & Warren, R.C. 1994. The concept of race and health status in America. *Public Health Reports*, 109(1), 26–41.

Williams, G., & Mawdsley, E. 2006. Postcolonial environmental justice: Government and governance in India. *Geoforum*, 37(5), 660–670.

Wolsink, M. 2006. Invalid theory impedes our understanding: A critique on the persistence of the language of NIMBY. *Transactions of the Institute of British Geographers*, 31(1), 85–91.

9 Evaluating environmental risks to health

Introduction

An implicit theme throughout this book is that social values play a role in all aspects of environmental health risks and their management. This chapter more thoroughly discusses how values are involved in deciding whether to be concerned and take action on environmental health threats and how society and science come together to make such decisions. Human values are infused in one way or another at every stage of environmental assessment and management. Even in Beck's risk society (Chapter 5), we tend not to think about risk on a day-to-day basis, and agencies responsible for managing risks are inclined to disconnect the science of risk assessment from the political management of those risks. Such decoupling of science and management is becoming increasingly difficult as lay publics more thoroughly comprehend the limits and uncertainties that are inherent in the science of toxicology and epidemiology and allied fields used to assess the likelihood of exposure and harms to human health. This chapter reviews the basic processes for assessing the health risk from (i) a single substance – risk assessment – and (ii) a facility or major project – environmental assessment with health impact assessment. However, both of these are tightly intertwined to assist in making decisions about what substances and projects are deemed safe enough to proceed.

Risk assessment

Human health risk assessment

Human health risk assessment concerns the likely adverse health effects from a substance including pathways of exposure, the dose-response relationship for the substance, particularly susceptible populations, and exposure conditions that might put certain groups at particular risk. As shown in Chapter 1 the dose-response curve from toxicological studies is used to determine the safest doses, like the dose (level) at which there is no observed adverse effect (NOAEL) or the lowest dose at which there is an observable effect (LOAEL). Nevertheless, the NOAEL and LOAEL are merely the start point for discussions about safe levels for humans (Moeller, 2011). The point of this section is not to review risk assessment, the dose-response curve or NOEL and LOAEL, covered in earlier chapters, but to identify potentially value-laden decision points within those processes.

Human health risk assessment often goes hand in hand with ecological assessment, and though this chapter focuses only on human health it is important to recognize that ecological risk assessment and human health risk assessment can intersect considerably, particularly for the large array of substances that have direct and simultaneous impacts on human health and other organisms. Controlling the release of such substances warrants particular attention since impacts are potentially very widespread. For instance, substances that *bioaccumulate* (the intake of substance exceeds and organism's capacity to excrete it) or *biomagnify* (relative concentrations of a substance increase up food chains and across food webs) tend to have both ecosystem and human health implications, with humans typically found in prominent positions in food webs. The next section details the risk assessment process for a single substance.

The risk assessment process and social values

New chemicals brought to market in most developed industrialized countries require risk assessment to assure consumers that the chemical is indeed safe – but 'safe' is a relative term. Various government regulatory agencies such as those which deal with food, drugs, pesticides, health and environment oversee that the threat to human health is within a tolerable range. The goal of these agencies is not absolute safety but, instead, risk minimization – a value judgement built into the system of risk assessment. In Canada, for example, the Pesticide Management Regulatory Agency (2014) identifies risk minimization as a guiding principle in the following: "Pesticides are stringently regulated in Canada to ensure they pose minimal risk to human health and the environment". This presumes that there are certain costs – for example, risk of health impacts – that we as a society are willing to accept or, at least in the context of a risk society, that we are *assumed* to be willing to accept. Regardless of whether we personally subscribe to such a value judgement, there is a complex system in place in many countries to fulfil the social/regulatory goal of 'minimal risk'. However, there is rarely public discussion on how minimal is minimal enough.

Values seep in at all stages of the risk assessment process (see Figure 1.1). The U.S. Environmental Protection Agency (EPA) defines *risk assessment* as a process used to "characterize the nature and magnitude of health risks to humans (e.g., residents, workers, recreational visitors) and ecological receptors (e.g., birds, fish, wildlife) from chemical contaminants and other stressors that may be present in the environment" (EPA, 2014a). This process is often more precisely termed 'quantitative risk assessment' in that it is designed to simplify the problem of risk down to specific quantities (Lerche and Glaesser, 2006; Simon, 2014). Those quantities come in two forms: first, the *probability* of a negative consequence occurring and second, the *magnitude* of that consequence. For some risk analysts then, the calculation of adverse effects is crudely defined as the probability of it happening multiplied by the magnitude of effect. Both quantities can be seriously fraught with difficulties, as reducing inherently complex systems to numbers involves simplifications and definitions upon which various stakeholders involved may not agree. Nevertheless, it is this quantified assessment that is used by risk managers to decide how to prevent or minimize harm by either decreasing the likelihood that negative impacts will happen or decreasing the depth and extent of the damage or all of the above. Risk managers are generally viewed as experts working in government or industry, but as the EPA themselves suggest, experts nowadays increasingly include highly informed laypeople who have a stake in a particular development or environmental threat. The most common case of laypeople being, in effect, 'risk managers' is when they get involved as parties in an environmental assessment process to ensure a new development is kept to a high standard of safety. For example, Brown describes what he calls *popular epidemiology*, in which communities work with academics to collect their own population health data. Though the academics may guide design and analysis, the process is meant to empower communities in the decision-making process. Yet in Brown's case of Woburn, Massachusetts, this heavy community involvement did not happen until after a contamination event and childhood leukemia cluster had already developed. How developments are assessed ahead of approvals is discussed later in the chapter.

Risk assessment is often represented as a step-by-step process as in Figure 1.1. In broad terms the flow chart shows that *risk assessment* is often distinguished from *risk management*, with the former comprising a number of steps oriented towards quantitative science and/or social science while the 'management' component involves more decision making which invokes social values about how to proceed. The broad academic discipline that deals with all of these aspects is often called risk analysis with the Society for Risk Analysis (SRA.org) being the most prominent professional society. If a hazard is defined as a substance that has the potential to adversely impact humans or other organisms in an ecosystem, *hazard identification* is a largely value-oriented social/institutional process of deciding that a substance should be assessed in one way or another. *Hazard characterization* is the process of collecting scientific information on the chemicals involved to provide quantitative and qualitative descriptions of their potential to do harm. This is where the toxicological and epidemiological evidence described in Chapter 1 is brought to bear. Toxicological studies, in particular cause–effect dose-response relationships, come together with environmental exposure assessment to determine the overall level of threat to humans. When hazard characterization determines that the substances involved are associated with extremely low probability and low magnitude of harm, the process can potentially stop in support of a do-nothing scenario to minimizing human exposure as the substance is allowed to go to market. If there is even a small potential threat the process proceeds to *environmental exposure assessment*, which determines how various organisms, particularly humans, are potentially exposed to the substance (e.g., through air land or water; over long or short distances).

Though the 'risk assessment' stages in Figure 1.1 may suggest that these are largely technical and mechanical processes they are nevertheless value based to a degree. For example, Chapter 1 outlines how a dose-response curve requires some assumptions that are value based. That is, since we only know what happens at the very high doses used to test the effects of chemicals on animals over relatively short periods of time, scientists must make judgements about the effect of much lower doses of those same chemicals on humans over much longer time periods (Moeller, 2011). That curve extension from what is known at high doses down to where it crosses the response axis at lower doses (Figure 3.3) is a value judgement.

Risk management

Risk management is focused on providing a response that is proportional to the risk posed by a hazard. This involves decision making at a regulatory level that sometimes spills into broader political and societal debate. That is, this aspect of environmental health risk analysis is more bluntly and visibly value laden. Risk management is much more involved with politics, economics and society generally than is the risk assessment process, but as Beck's theory of a risk society suggests in Chapter 5, since laypeople are increasingly questioning the science behind risk assessment itself, distinguishing a value-laden (bottom part of Figure 1.1) and ostensibly value-free (top part of Figure 1.1) component to overall environmental risk analysis is increasingly blurry. Thus, very similar risk characterizations can lead to very dissimilar regulatory responses. For example, one country might choose to severely restrict or ban the same substance that another country may decide to allow to be conditionally marketed. Examples are explored further in the chapter, following a more detailed discussion of risk evaluation, emission and exposure control, risk monitoring and risk communication.

Risk evaluation is a process that bridges the laboratory scientific and exposure assessment processes with the more management-oriented aspects of decision making. At its core this process is about weighing the costs and benefits of the substance involved, which can include both quantitative and qualitative information and is sometimes called cost/benefit or risk/benefit analysis. Thus, while the steps mentioned earlier are about assessing the toxicity of the substance, how it is likely to get into the body and then into the organs/systems that might be negatively impacted, risk evaluation adds to the process the consideration of benefits. For example, a proposed new pesticide for controlling mosquitos is likely to have some sort of benefit in reducing exposure to vector-borne diseases like malaria and West Nile virus. The quantitative degree to which mosquito populations and those diseases may be proportionally reduced might be combined with quantitative or qualitative information on the perceptions of the diseases themselves – for example, the degree to which they are dreaded – in the areas where the pesticide is to be licenced. Thus, it is far more difficult to justify an even remotely hazardous pesticide for so-called cosmetic control of weeds or other biting insects compared to justifying it for direct health benefits. DDT is a well-known pesticide that is still controversial because it is both dreaded as a potential ecological threat and potential human carcinogen and revered as an inexpensive mosquito killer to control malaria and other mosquito-borne illnesses. Such cases in which benefits are high in certain contexts tend to evoke the greatest public controversy (López-Carrillo et al., 1996; O'Shaughnessy, 2008) whereby for example the Stockholm Convention of the United Nations signed a treaty banning several pesticides including DDT. Yet there is a caveat built in that allows for the specific use of DDT to control mosquitos in malaria-prone areas (Stockholm Convention, 2014). In effect, DDT is therefore fully banned in places in the Global North but not necessarily in equatorial regions prone to the disease. In terms of risk/benefit analysis the World Health Organization's (2011, 1) position is as follows: "The Convention has given an exemption for the production and public health use of DDT for indoor application to vector-borne diseases, mainly because of the absence of equally effective and efficient alternatives."

It is worth reinforcing that there are two key factors that contribute to this decision: (i) sufficient lack of clear epidemiologic evidence that there are serious human health impacts at typical DDT exposure levels in malaria endemic countries and (ii) DDT is in low cost for controlling malaria. There is a complex social history to DDT as well. Some argue that the negative human health effects of DDT came onto the scene long after the substance was banned for its deleterious effects on non-human components of ecosystems (e.g., birds and fish; Carson, 2002). This, some argue, leaves room for speculation about whether it ever would have been banned if current systems for risk analysis had been used, that is, a system that considers human health benefits along with the risks (Edwards, 2004). Indeed, Conis (2010) argues that at the root of the debate is a difference of opinion on worldviews and values. Some highly value, for example, sustaining overall ecosystem species diversity, while others place a much higher priority on human health. This point will be revisited in our discussion of cultural worldviews and rationalities towards nature in what follows.

If a substance is deemed toxic enough to warrant mitigation, risk assessment will recommend or take account of *emission and exposure control* measures. These are measures that, if implemented, would reduce the overall risk to human health by either preventing the substance from going where it is not supposed to and/or preventing people from being exposed in other ways. In the example of a new pesticide, this could include detailed labelling that describes how the pesticide is to be applied. It might be for use on standing water only if it degrades quickly there. DDT, for example, can be very effective for controlling mosquitos when sprayed on the walls of homes. Because it is persistent, it can be effective for several months. However, under the Stockholm Convention, an explicit control measure is to prevent its use on crops since it indiscriminately kills several sorts of insects and may lead to mosquitos building resistance to the toxin.

Once a substance has been approved for use there may be provisions put in place to conduct *risk monitoring* to ensure that the threats are no higher than predicted – for example, that exposure levels are not higher than anticipated. This is why the WHO constantly considers new data on DDT, to detect if there is human epidemiologic evidence of deleterious effects (since we already know DDT is linked to cancer in lab animals) that warrants reassessing emission and exposure control measures or banning DDT altogether. Given the benefits of DDT, such a ban might entail making judgements about substituting a less toxic alternative. Unfortunately, it is very difficult to monitor all substances deemed safe enough for use. By some estimates we already have about 70,000 human made chemicals, and about 200 to 1,000 new ones are introduced each year (Moeller, 2011). The cost of monitoring all of these chemicals through epidemiologic studies is cost prohibitive and would require an even more dramatic shift in resources globally than Beck, Giddens and others suggest is already happening as we transition to a 'risk society'. Thus, the decision to conduct limited monitoring is a social value decision at the broadest scale.

Risk communication concerns the exchange of information between interested stakeholders, one of whom is the general public(s). Several lessons have been learned over recent decades about the perils of thinking of risk communication as simply a process of informing the public, whereby expert scientists and environment and health managers decide what messages to convey to the public and, in turn, expect the public to adjust their behaviours accordingly. Even when specific behaviours are identified as essential – such as wearing a respirator when applying a pesticide, or not smoking – risk information is ignored for social, cultural and political reasons. Testing for radon in homes is a good example of an environment and health problem with a fairly well-known dose-response relationship with lung cancer but whose public risk communication campaigns have had limited success (Poortinga et al., 2011; Johnson and Smith, 2013). For example, in the Poortinga et al. study of England and Wales only 19% of their entire sample had completed an in-home test for radon despite the fact these researchers specifically sampled in areas known to be higher risk.

On the other side of the coin of risk communication, there are substances that risk managers may consider minimal risk, but the public may demand that the substance be more stringently regulated. One example is Rachel Carson's *Silent Spring* (2002) and the associated backlash against pesticides that led to the banning of DDT in the United States, Canada and elsewhere. Leiss (1996) traces how risk communication has evolved over time to better incorporate the complex relationship that has developed between various publics and environmental health hazards. The complexity is encapsulated in the risk society thesis from Chapter 5 with the public becoming increasingly concerned about large-scale, catastrophic risks while they are simultaneously sceptical that experts, including scientists themselves, will make the best decisions based on the available science. Mistrust in experts is a fundamental problem for risk communicators. Society has moved well beyond a time when the focus was mainly on getting the numbers right and simply treating risk communication as an afterthought. Among other things this is due to the equivocal nature of risk assessment and the value-laden nature of the courses of action that are appropriate – a central theme of this book.

Another problem in the history of risk communication is that much of it has been based on a *knowledge gap/deficit model* of lay understandings of risk. This model assumes that the reason people fail to act to (a) take precautions to minimize risks deemed unreasonably high by experts (e.g., protective equipment when spraying pesticides, home radon testing) or (b) reduce their concern over hazards many experts agree are sufficiently minimal (e.g., fluoridation of water) is because people simply do not have access to or understand risk assessments. The irony in the case of scenario (b) is that the people most concerned to regulate certain hazards tend to also be most informed *because* of their concern about threats to health. Hansen et al. (2003) trace this idea in relation to food risk concerns, particularly genetically modified foods. They argue that heightened concern about such risks has more to do with social context, a sign of the times, than it does with people failing to understand how food systems threaten their health. For example, though *E. coli* is arguably a more real and immediate danger to health than genetically modified organisms, the scale of potential harms and the values associated with human impacts on ecosystems motivate action against GMO threats. We return to these ideas in our discussion of risk amplification and worldviews towards nature.

One of the themes developed by Leiss (1996) is that the more recent phases of risk communication are characterized by an exchange of information. He suggests that information should not flow only one way, from experts/scientists to lay publics; instead, there should be open communication in both directions. This idea has grown up in the information age of the Internet, which readily facilitates fairly instantaneous two-way communication of all kinds (chat, forums, bulletin boards, blogs, news media commenting systems). One of the goals of this two-way exchange is to regain trust between lay publics and the institutions entrusted with managing environmental health threats. For example, Health Canada (2014a) has enshrined this two-way exchange of risk information in their recent framework for strategic risk communications, which is defined as "any *exchange* of information concerning the existence, nature, form, severity or acceptability of health or environmental risks" (*emphasis* added). Unfortunately, trust is far easier to lose than it is to gain – something called the *asymmetry of trust* (Slovic, 1993; Poortinga and Pidgeon, 2004) – and the effort to regain trust through such an approach to risk communication is often hampered by momentary lapses back into outmoded ways of thinking, like the knowledge deficit model (Wynne, 2006).

Environmental risk assessment of a project or cleanup site

So far, the focus has been on the assessment of single substances, but risk assessment often refers to environmental impacts assessments, or EIAs (or simply environmental assessments – EAs) of projects. Though EAs generally involve a wide range of ecosystem, social and economic impacts, increasing there are specific health impact assessment (HIA) components included (Ozonoff, 1994; Lerer, 1999). While health risk assessment was designed mainly for the evaluation of single chemicals, environmental impact assessments involve physical sites and projects (e.g., pulp mill, hazardous waste treatment centre, oil pipeline) with numerous chemicals and potential pathways of exposure, which dramatically increase complexity. In effect an assessment of, for example, a brownfield site (contaminated land previously used for industrial purposes) might require attention to the individual risk assessments of numerous chemicals involved at the site (e.g., creosote and diesel fuel at an old rail yard). Yet there are at least three parallels between the two processes. First, hazard identification health risk assessment is similar to the scoping stage of environmental assessment since both involve identifying which things will be assessed and, by extension, which things will not be assessed. Whereas hazard identification is relatively straightforward – a chemical is identified for assessment – scoping is more involved for EAs and potentially includes a range of social values.

Second, EA analysis is generally conducted by consultants on behalf of the proponent to the undertaking (e.g., a private company or a government agency), including taking baseline samples from the ecosystem and potentially affected human populations. This is similar to the *risk characterization* stage of risk assessment. Whereas risk characterizations within chemical risk assessments are based on laboratory toxicological work and general population exposure scenarios, EA analysis narrows the context to a specific population near a specific proposed facility or other potentially hazardous undertaking. Third, the final EA report and associated hearings, if hearings even happen, are roughly parallel to the risk evaluation stage of risk management, as both are the stages at which a decision is made about the level of threat to human and other populations. Fourth, the approvals and follow-up stage of EA can be quite similar to the emission and exposure controls and monitoring stages of risk management. That is, EA often requires certain conditions for the proponent to meet. Those conditions usually involve monitoring certain ecosystem components to detect environmental degradation and the potential for human health threats. Whereas risk assessment monitoring for chemicals will involve very broad epidemiologic studies in the general population, monitoring for an EA will tend to only focus on the areas surrounding the specific development.

Though both health impact assessment within EA and health risk assessment are value-laden processes, EA ostensibly has public involvement at all stages of the process at which the public can have their say. For example, at the scoping stage local populations may request that specific maternal and birth outcomes be analyzed in the EA, whereas the analysis of particular chemicals in a risk analysis is more routinized and general, though vulnerable populations (e.g., infants, elderly adults) are more commonly being used as the basis for determining acceptable intake values. Likewise the public can formally participate in hearings about the facility if the EA process is of sufficient scope to require such public involvement. For example, the now famous Mackenzie Valley Pipeline environmental assessment which spawned the Berger Inquiry environmental assessment report (Berger, 1977) took years of public consultation to come to the conclusion that a major gas pipeline proposed in the 1970s along the Mackenzie Valley in Canada's Yukon and Northwest Territories should not be built. In fact, because of the extensiveness of public consultation that particular EA process is often held up as a gold standard for public involvement. Though public participation is an ideal of the EA process, it has also been heavily criticized for being too technically complex, exclusionary, lacking in clarity of focus and inadequately funded (Stern and Dietz, 2008). So the degree to which public values are adequately incorporated in the EA remains debated, particularly when a power imbalance between proponents and the public is left unchecked (Fischer, 2003).

Though risk communication is not an explicit step identified in environmental assessment it is something that may happen incidentally throughout the process through public participation. Similarly, a condition of approval of a project based on an EA report may be to conduct ongoing risk communication in the community – for example, through a public liaison committee and website – to ensure they are kept up to date with and involved in decisions surrounding the facility. Though a full description of HIA and EA is beyond the scope of this book, a brief example will put the process into focus.

Consider the example of a proposed waste-to-energy (WTE) incinerator. Before the facility can be approved it must undergo an EA which will typically include some form of risk assessment of potential harm to the health of the local community – that is, an HIA. For the HIA, the proponent and directly affected parties to the incinerator will want measured anything about the facility that may impact human health of those working at and living around the facility, including transportation of materials into the facility (e.g., road collisions), the processes that take place in the facility (e.g., diesel fuel spills) and substances that leave the facility (e.g., air pollution from the stacks). What this measurement typically amounts to is assessing pathways of exposure to set a baseline for what the environment is like prior to the facility and how harmful substances may get into human bodies. Thus, any monitoring that happens once the incinerator is operational will be analyzed for elevation above that baseline. For example, three major categories of toxic chemicals released from incineration facilities are polycyclic aromatic hydrocarbons (PAHs), chlorinated hydrocarbons (CHCs; e.g., dioxin) and toxic metals (e.g., chromium VI). All of these materials may be transported on fine particulate matter (PM, e.g., PM2.5), thus considerable attention is paid to the amount of such materials, particularly PM2.5 released from incineration facilities (Rushton, 2003; Cormier et al., 2006).

The health impacts that may potentially result from these substances include cancers, cardiovascular disease, respiratory ailments, neurological problems and low birth weight. For example, blood samples may be taken from workers at the onset of their job to set the baseline for these substances and allow for jobsite mitigation (e.g., shift to duties in lower-exposure areas). Chapter 1 outlines how such health impacts are usually determined through toxicological assessments in a laboratory environment using animals as proxies for humans. Thus, the risk assessor working in the context of conducting an HIA within a larger EA is challenged with putting together the information about how the facility will operate with the laboratory toxicological knowledge. The decision about whether to proceed is based on the sum total of all types of impacts, taking account of mitigation efforts to prevent short- and long-term harm. Since toxicology and epidemiology are sciences with inherent but known design limitations and uncertainties, such uncertainties are added onto the uncertainties associated with actual facility release information from a hypothetical facility. Yet EA and HIA can involve complex politics connected to the science itself, whereby proponents may be supported by an entire industry that systematically attacks scientists whose work is critical of industry science. Eyles and Fried (2011) highlight such tactics in the context of the nuclear industry in Canada putting forward an environmental impact statement for a nuclear waste repository in Kincardine, Ontario. They point out the efforts by the nuclear industry to discredit science that disagrees with the idea that deep geological disposal poses sufficiently minimal risk.

International differences in risk assessments

The value-laden nature of risk decisions is underscored by national differences in risk assessments, which are often ostensibly based on the same publicly available scientific study evidence. When guidelines and recommendations for exposure limits differ by nation, this suggests that the decision-making processes involved in risk management are the culprits. This must be put in the context, however, that even scientists can disagree on the meaning of the science itself – as highlighted in Chapter 4.

For example, though the current guidelines for household radon exposure in Canada (200 Bq/m^3) now roughly align with those in the United States (150 Bq/m^3), from 1988 to 2007 the recommended allowable exposure level was four times higher and less stringent at 800 Bq/m^3 (Health Canada, 2014b). The drop from 800 down to 200 need not be motivated by new science showing greater risk; rather, an adjustment in the safety (uncertainty) factors built into the risk analysis can account for the adjustment. One of the reasons for the delay in the reduction of this guideline is that naturally occurring background radon levels are quite high in some parts of Canada (e.g., Nova Scotia, interior British Columbia), and cost benefit analysis can easily lead to the conclusion that widespread public health measures will not be cost effective relative to investing those same funds in other parts of the health system (Létourneau et al., 1992). This represents an expert societal value judgement about the allocation of resources. More recently though, the United States had demonstrated the effectiveness of putting the onus on home builders to mitigate where necessary in new home construction and homeowners to test and mitigate at their own discretion (and expense) in existing homes. In effect, the United States and other jurisdictions tested out a relatively cost-effective mitigation scheme which Canada also adopted in support of their new lower recommended exposure guideline level.

Even within the European Union, where there is considerable harmonization of policy, there can be seemingly stark differences in approaches to setting acceptable standards for industrial project risk. Ale (2005), for example, traces the very different legal and other institutional contexts for the Netherlands compared to the UK, where the former is based on a Napoleonic legal system where the law sets the standard, whereas in UK common law in many ways, only 'starts the conversation' about risk. As with the case of radon, though, Ale also traces how the two seemingly different histories have led to many parallel risk outcomes. These outcome similarities are despite the fact that they each took very unique paths to arrive at very similar guiding criteria. For example, whereas the Netherlands established a maximum allowable individual risk of *death* of 10^{-6}/year for an undertaking the UK settled on a maximum individual risk of *exposure* to a lethal dose (with no specific reference to health or death) of 10^{-5}/year. These criteria have led to very similar standards and ultimately tolerable daily intake doses on a range of substances. One of the key points of discussion that guided these decisions was to account for situations in which catastrophic losses may result. This may be one of the key reasons the different guiding criteria have led to the similar standards.

Both the radon and industrial hazard standards examples seem to suggest that despite very dissimilar social and institutional contexts national standards for risk need not differ considerably. This may mean that societal values regarding, for example, the high valence of catastrophic losses and the value of human life, are similar across some national contexts. Indeed, such a theme is generally supported by comparative international research on risk perceptions where, with a few exceptions, residents of various countries replicate the same dread/unknown factor space that was produced in those early psychometric studies based on U.S. populations (see Chapter 4; Renn and Rohrmann, 2000). What may differ though is the regulatory context. For example, more nuclear plant electricity is generated in France because of a relatively technocratic, top-down regulatory system, not because the average citizen fears the risk of nuclear catastrophe any less (Slovic, 1993). Likewise Teräväinen et al. (2011) use discourse analysis to argue that the nature of state support for new-build nuclear power can vary considerably, with France taking a 'government-knows-best' approach, Finland holding a 'technology-and-industry-know-best' attitude with the UK leaning more towards a 'market-knows-best' approach.

Similarly, Wiener and Rogers (2002) suggest that despite the fact that the U.S. and the EU have very tight regulations on their food systems, the U.S. has reacted in a more precautionary way towards mad cow disease in beef, while the EU is more stringent in regulating hormones in beef. Thus, regulatory differences will vary according to social and political context, including the interplay among the state, the market and civil society. Despite these policy differences the underlying perceptions of risk from the average person may remain very similar across several nations (Renn and Rohrmann, 2000). In fact, Jasanoff (2005) argues that with global changes in political systems such as a general weakening of state sovereignty and a rethinking of governance generally we may see greater international and even intranational differences in acceptable standards for various technologies and substances. She suggests that

> the 'old' politics of modernity – with its core values of rationality, objectivity, universalism, centralization, and efficiency – is confronting, and possibly yielding to, a 'new' politics of pluralism, localism, irreducible ambiguity, and aestheticism in matters of lifestyle and taste.
>
> (Jasanoff, 2005, p. 14)

She argues that politics and culture – which are often most keenly felt at a subnational level – may prove to be as important as science and the state in decision making on risks.

Challenges of risk assessment

There are a number of challenges and critiques of risk assessment, environmental assessment and health risk assessment concerning both the involvement of stakeholders and safeguarding health. Some of these challenges are specific to the assessment of projects as opposed to chemicals, but many of them are common to both since project assessment ultimately relies heavily on toxicological and epidemiological evidence. These challenges may be roughly categorized into those that stem from (i) the limitations of the sciences involved (e.g., exposure assessment and delayed effects) and (ii) those that stem from the social/institutional structure of risk management (e.g., defining hazard, cumulative effects assessment, psychosocial health, winner take all vs. win-win, decision authority, feedbacks from monitoring).

Exposure assessment

Exposure assessment is covered in Chapter 1, while the purpose here is to highlight limitations that open the science of risk assessment to debate. The challenge of exposure assessment is that while the doses of chemicals are closely controlled in laboratory environments (toxicology), measuring doses in the public (epidemiology) is constrained by the social and institutional contexts in which research can be conducted in the real world. Though the ethics of experimenting on animals is debated, there is a tacit assumption that it is unethical to experimentally expose humans to toxic chemicals. If we are to understand how the health effects of chemicals on non-human animals in the lab translate into health effects in the field, we must search out places where humans are either already exposed or possibly exposed. This seems fundamentally illogical since humans presumably should not be exposed until a chemical has been assessed for safety. Yet as indicated, in this chapter there are so many new chemicals produced each year that each may receive only cursory risk assessments before going to market. These assessments may be as simple as test tube analysis or by drawing parallels from studies done on other very similar chemicals. One of the reasons this is allowed to happen is that in most instances mitigation and safety efforts in industry in particular result in very low exposures to humans. Thus this system will eventually let some substances slip through without adequate warning of as-yet-unknown harmful effects – for example, thalidomide (see Chapter 10). When there is reason to be concerned, long-term toxicological studies are indeed carried out to determine the impact on animals. This brings us back to the original problem of comparing high exposures in the lab over short periods of time with low exposures in the real world over much longer periods of time.

Even with human exposures available in the real world to study, risk assessors often do not have direct exposure information – the ideal being information about how much of the chemical reaches so-called target organs, the bodily sites where the chemicals are most biologically plausible to cause negative health effects. It is not practical or cost effective in epidemiologic studies to take tissue, blood or other human samples in most cases, which leaves indirect proxies of exposure as the best available measures. Working at a facility that already handles a certain chemical (e.g., a DDT manufacturer) is commonly used as a proxy. Thus, epidemiologists often rely on worker studies because those workers tend to be the population with the highest exposure to various chemicals. If there are no health impacts detected in this high-exposure group, then the rest of the public is far less likely to be negatively impacted at the lower doses they would experience. For example, an epidemiological meta-analysis of *formaldehyde* in industries such as embalming, pathology and various chemicals industry firms revealed a slightly elevated risk of leukemia in some cases but a reduced relative risk at the highest doses in one particularly strong study. The authors suggest then that overall there is not enough consistent evidence to suggest that formaldehyde exposure in these workers causes leukemia (Collins and Lineker, 2004). However, the latter is a value judgement based on what to do in the face of inconsistent and limited evidence. Such findings might just as easily support the status quo as they could prompt a review of further management of formaldehyde if risk managers were inclined to be precautionary. In the absence of worker studies exposure is generally modelled based on some combination of proximity to a proposed facility/chemical and likely routes of exposure. For example, dioxins released from an incinerator stack can be modelled according to local conditions in an air plume model. Sometimes even plume modelling is not feasible given available resources, in which case circular exposure zones are often delimited around the facility. Thus, the practicalities and costs of measurement tend to lead to relatively poor exposure measurement in practice – at least from a scientific point of view.

Delayed effects

Chapter 7 goes into detail on delayed effects, so the purpose here is simply to link some of the themes to the more practical challenges of risk assessment and management. There are a number of temporal issues to consider in environmental health impact risk assessment which link back to some of the known limitations of exposure assessment. While some effects are acute when there is unexpected release of large amounts of a substance in a short period of time – on the order of hours – a more common scenario is the release of small amounts of a substance over long periods of time – on the order of years. Acute effects are somewhat easier to assess because they more closely mimic laboratory situations. For example, lethal doses are administered (i.e., LD50 studies) to determine the level that is expected to kill laboratory animals. The accidental high-volume releases at Bhopal, India, of methyl isocyanate (MIC) in 1984 and in Seveso, Italy, of dioxin (TCDD) in 1976 are examples of high-level acute exposures, which gave risk managers further information about the toxicity of these substances in humans. MIC at Bhopal killed thousands, while acute TCDD exposure was attributed moreso to non-lethal conditions like chloracne. However as Chapter 7 details, the long-term health impacts are far more complex to measure where, for example, one or two decades may pass between exposure and outcomes such as cancer and even birth defects. Simply tracking the population that was exposed is difficult, especially since those who survive initial exposure may prefer to move away to areas perceived to be safer from a repeat disaster. In both cases the long-term impacts have been significant, with cancers leading to death being among the most serious (Eckerman, 2005; Consonni et al., 2008). For example, in Seveso, Consonni et al. found elevation in deaths from lymphatic cancer with risk ratios relative to controls ranging from 1.6 to 5.0 in the zones away from the facility where the TCDD was released. In such a situation, where uncertainty is high and the best epidemiological evidence is not available until it is too late, worry, anxiety and depression are among the psychological impacts that are experienced locally. Risk assessment generally does not account for psychosocial impacts, and this issue is discussed further in what follows.

What counts as hazard?

Though health impact assessments ostensibly evaluate the impacts of specific chemicals or the overall safety of operations associated with a facility there may be other aspects of both chemicals and facilities that impact stakeholders. Chapter 4 describes how 'dread' of a chemical can influence how it is perceived and how much people desire it to be further regulated. Thus, while hypothetical chemicals A and B may both be assessed to represent a relatively low risk to the general population, if chemical A is highly dreaded compared to chemical B, people may demand more stringent rules or an outright ban – DDT is one example. In this sense it is more than just the chemical that represents the hazard; it is the social, cultural, political and other processes that ratchet up or ratchet down public sentiment in support of or against various projects and substances.

Such situations are explained by the social amplification (and attenuation) of risk framework (SARF) developed by Kasperson and colleagues (Kasperson et al., 1988; Pidgeon et al., 2003; Figure 9.1). This framework is meant to describe both (i) high concern about hazards deemed by experts to be relatively low risk and (ii) low concern about hazards deemed by experts to be relatively high. DDT would fall into the amplification category for many people. The framework brings to bear a range of risk research (psychometric, cultural, risk communication and the media) to understand not only how the public responds to hazards but the secondary impacts those responses might have in the broader social, political and economic systems. How people interact with the media on hazard risk is of particular interest. The media are prone to report the more dramatic aspects of any story, including environment and health impacts (Pidgeon et al., 2003). These impacts may or may not be covered under the auspices of an environmental assessment as described, particularly negative psychosocial impacts on communities. Further, the framework explains processes in the stigmatization of a technology or facility and all related facilities from social amplification of

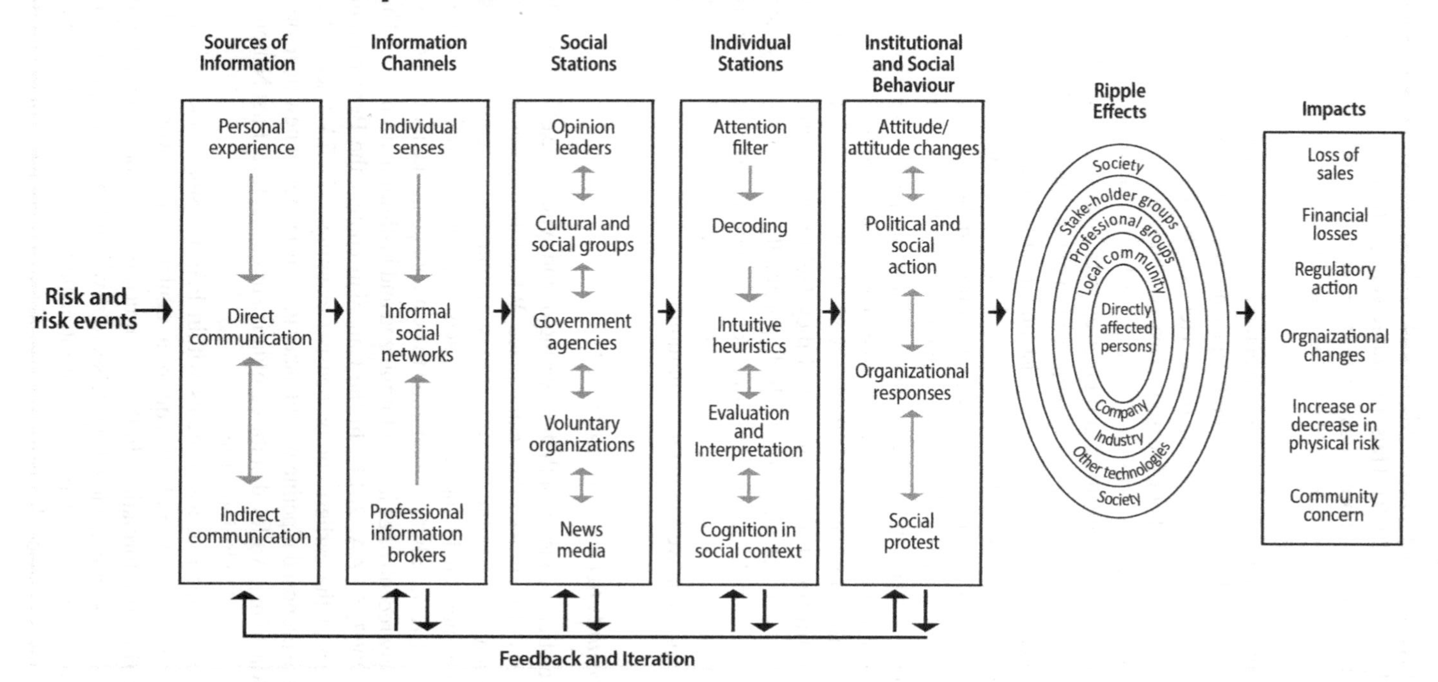

Figure 9.1 The social amplification and attenuation of risk framework

Source: Pidgeon et al. (2003), reproduced with the kind permission of Cambridge University Press

risk. Mad cow disease is one example of how amplification of concern in the media resulted in stigmatization of the beef industry in countries like Canada (Lewis and Tyshenko, 2009). Thus, the framework particularly brings into focus the unintended impacts of how social (family, work and friends) and institutional (e.g., media, risk assessment agencies) systems interact to produce a range of unintended negative impacts.

The definition of hazard might also be extended to more than just the chemicals emitted from a facility to include the EA process itself. For example, participating in an environmental assessment process tends to be stressful for local directly affected people for a number of reasons (Deacon and Baxter, 2013). Environmental assessment is part of day-to-day business for facility proponents – they fully expect to hire lawyers, scientists and other experts – yet this is not the case for the average resident. This lack of experience is a conundrum for all stakeholders involved, since resident groups often insist, and most EA legislation requires, that local residents be involved in the decision making involved in the environmental assessment process. While consultants, regulators and the proponent are paid for their efforts and time, locals – despite some availability of intervenor funding (Jeffery, 2002) – and other stakeholders generally are paid minimally or not at all, so this can put a strain on the resources locals can commit to involvement. Further, once local residents and other concerned stakeholders become involved, they may be frustrated by the process if they feel their concerns are not being addressed (O'Leary et al., 2004), which potentially compounds psychosocial impacts. Expanding the scope of what we consider impacts to include the process itself may go some way to understanding how EA processes might be changed to minimize impacts on well-being in communities.

Cumulative and effects assessment (CEA)

Cumulative effects concern both the layering of and synergistic effects of releases from all types of facilities in the same place. What this requires of EA is that all past, present and reasonable future facilities are taken into account when assessing environmental (including health) impacts of a new undertaking. Some have argued that CEA was always understood to be a core component of good EAs, but only relatively recently has it been formalized and institutionalized (e.g., Therivel and Ross, 2007; CEAA, 2014). In the Canadian context the Berger Inquiry is probably one of the earliest comprehensive examples of CEA since it became clear that the proposed pipeline would usher in such a range of new developments along the Mackenzie Valley that the cumulative threats to local First Nations ways of life were judged to be too large.

CEA complicates risk assessments within EAs for single undertakings in a number of ways. For example, it requires a regulatory framework in which there is open sharing of information. In the past much of the information required for CEA may have been considered 'trade secrets' within highly competitive industries. It also requires that there be a clear rationale for scoping cumulative effects over both time and space. Knowing the spatial and temporal extent of cumulative effects is difficult to know a priori, but as time and space increase so too does the complexity of the assessment. Further, adding up releases from multiple facilities may be relatively simple compared to assessing the *synergistic effects* of chemicals on each other. For example, Kortenkamp (2007) finds that while individual endocrine disruptors (e.g., estrogenic, anti-androgenic) may not cause health effects at low doses, when they are combined at those same doses they can produce observable health effects. His recommendation that endocrine disruptors need to be analyzed in combination as well as individually greatly complicates toxicological and epidemiological analysis. In fact, the Agency for Toxic Substances and Disease Registry (ATSDR) argue that any sort of global analysis of chemical mixtures is not possible:

> Individual testing of the endless number of potential combinations is virtually impossible. Even if cost were not considered, the number of animals required to perform statistically relevant toxicity tests with multiple doses for multiple exposure periods would be prohibitive.
>
> (ATSDR, 2015, p. 65)

The challenge then is to identify when it is necessary to assess chemical mixtures on a case-by-case basis.

Challenges such as these have led to a relatively poor track record of CEA in places like Canada and the UK (Cooper and Sheate, 2002; Therivel and Ross, 2007). For example, in a study of 12 years of EA statements in the UK, Cooper and Sheate (2002) conclude that only 48% even mention the term 'cumulative effects' while only 19% discuss cumulative effects in any appreciable detail. Therivel and Ross (2007) more critically examine the reasons for the poor performance in assessing cumulative effects, concluding that an inadequate methodology for scoping of the *spatial* scale of impacts is at the heart of the problems with CEA in Canada and the UK. While CEA is, in theory, valuable and increases the legitimacy of EAs, more work needs to be done to find a balance between the years-long assessment that was required for the Berger Inquiry and the much shorter timeframes expected for completing EAs today.

Psychosocial health

Often one of the tacit assumptions of health impacts assessment is that physical health, particularly disease symptoms, is the main concern rather than the broader definition of health that includes other forms of well-being (Ozonoff, 1994). If psychosocial impacts have to do with the interrelation between mental health and social interaction, assessing psychosocial health impacts also adds complexity to EA. For instance, if a facility or practice threatens a long-standing way of life in an area (e.g., subsistence hunting, farming) or the expectation of a certain way of life (e.g., a safe place to raise children or retire in peace and quiet) then there is more to impacts than likewise important nutritional changes in diet or safety of large trucks on roads – it has to do with changing a way of being (Barnes et al.,

2002; Baxter and Lee, 2004). For example, Crighton et al. (2003) trace the devastating social impacts on maritime ways of life from the loss of the Aral Sea due to industrial agricultural irrigation practices. Such impacts may manifest as stress, that is, insufficient coping mechanisms in the face of such threats (Lazarus and Folkman, 1984). In this sense psychosocial interactions can be positive or negative. It can be a negative impact on a community from exposure to a new facility and associated siting process, but it is also a force which potentially serves to minimize impacts. That is, social support is one of the most powerful ways to cope with any threat, including the psychological aspects of new facilities/events. Given that the process of being involved in public participation processes associated with EA is added on top of other sorts of worries such as contamination, and these can together amplify (or attenuate) risk within local social support networks, it is very difficult for risk assessors to predict and manage these types of impacts.

Winner takes all versus win-win

While the challenges listed can in some form or another be handled within the current institutional arrangements in most industrialized countries, a more fundamental critique of EA itself is that it is an inherently conflict-based system. Under most EA arrangements today there are generally two coalitions in the process: the proponent, along with other project supporters, and the opponents who do not want the undertaking to happen in the form proposed (see Chapter 2; Sabatier, 1993). This is, of course somewhat of a caricature, but late in the process it is difficult to avoid the fact that the proposed facility/activity either proceeds or it does not. This suggests that when a project does move forward, the proponents 'win', and when it does not the opponents do. This leaves the impression of winners and losers in each environmental 'battle' and that there are likewise risk winners (those who unexposed who benefit from the technology) and losers (those living close by who may or may not benefit; O'Brien, 2000; Durant et al., 2004).

Some suggest that antagonism over environmental issues is helpful if ultimately we assume that a single way of thinking will dominate in one way or another. For example, Schwarz and Thompson (1990) suggest that at the heart of every environmental decision is a valuable tension between the four worldviews from the cultural theory of risk (Chapter 5) as supported by associated assumptions about how the environment reacts to technology. They summarize the worldviews (rationalities) and the assumptions about nature represented as a ball on a surface. What happens to the ball (nature) when technology disturbs it is represented by the path the ball will take on the surface. For example, the individualist worldview is like a ball in a bowl – it favours the value of the market, materialism and open, unregulated competition and views nature as benign in the sense that nature is expected to relatively easily roll back to its original state after being subjected to an environmental hazard shock – no matter what happens to the ball it always rolls back to the same place in the bowl. The hierarchist worldview – represented by a ball on a wavy/corrugated surface – on the other hand means being more inclined to favour top-down decisions, accept the word of experts and prefer some regulation; combined with a view of nature as perverse – whereby *extreme* shocks can push nature over the edge towards destruction, rolling away and out of control. In this sense the ball representing the environment rolls to a new equilibrium each time it is pushed far enough. The hierarchist worldview supports regulation as a means to prevent those extreme shocks (like Bhopal, Seveso and Chernobyl) under the assumption that smaller contamination events are for the most part relatively harmless. The egalitarian worldview though favours collective processes, grassroots decision making and the sharing of power, supported by a view of nature as highly fragile or ephemeral, so that *any* shock to nature (not just extreme ones) is apt to push it towards serious uncontrollable decline – a ball precariously perched on a convex surface like a balloon or upside down ladle. Schwarz and Thompson argue that these assumptions are very resistant to change, leading to two important implications: (i) it is the tension among these three that has characterized most environmental decision making to date and (ii) consensus may be difficult because that suggests groups will have to adjust their worldview accordingly (see also Sabatier, 1993). That is, egalitarian environmental groups have kept both hierarchists and individualists in check by encouraging them to seek out technologies that are low scale and minimal impact, while hierarchists keep market-oriented individualists in check by legislating to control their market activities. For the time being the fatalists, who tend to be relatively poor and powerless within current decision-making structures, have less influence altogether – a ball on a table, it can roll anywhere at any time.

There may then be a tension between Schwarz and Thompson's view of how groups interact to make environmental decisions in society and a win-win system of decision making. Such a system tends to involve some form of compromise in which there is give and take in terms of addressing local concerns in particular. This idea is explored further in the section on alternatives to the current adversarial environmental assessment process. Yet even under the current adversarial system there are concerns about who gets to decide whether a project is safe enough to move forward, the issue discussed next.

Who decides and the precautionary principle

Most EA systems in the industrialized world are essentially designed like courts in which there is a single judge (e.g., a minister of environment or health) or jury (e.g., a panel of experts) that ultimately says yes or no to the final proposal with conditions. There is little room for consensus-based decision making. The egalitarians depicted by Schwarz and Thompson (see also Figure 5.2) fundamentally object to such an arrangement in that there is little sharing of power, so they are most likely to back out of the process entirely as a form of protest (O'Leary et al., 2004). Yet having any group backing out of the process puts the entire system in jeopardy, in this example by tilting too far towards coalitions that favour either the hierarchist or individualist worldview.

The precautionary principle is a standard for decision making that is often associated with the egalitarian worldview. Consistent with the egalitarian worldview, the precautionary principle suggests that when faced with uncertain scientific evidence about a substance or undertaking, the default position should be to regulate or not proceed further without changes. That is, the burden of proof is on the proponent of the substance or facility to show safety, presumably through heavy mitigation or further research. Thus the motto, "when in doubt, take it out (or don't do it)" might be a shorthand way to think about how being precautionary impacts risk management decisions. The Berger Commission decision not to move forward on the Mackenzie Valley (natural gas) Pipeline in the late 1970s in Canada is an early example of the precautionary approach in action. Whether those tasked with making decisions within risk assessment and EA processes take a precautionary approach varies widely, but compared to decades earlier it is more on the minds of decision makers than ever before (Raffensperger and Tickner, 1999; Foster et al., 2000).

Feedback – closing the loop

Figure 1.1 depicts an EA system that is rather linear, suggesting that there is a beginning and end to the process. The end ostensibly is marked by the denial or approval of a development that may be potentially hazardous to human health. Similarly, the risk assessment process is often depicted as mostly linear, but does include one feedback – between monitoring and emission and exposure control. What neither of these figures adequately accounts for is if, and how, learning happens within the system (Sabatier, 1993). The value of feedbacks is that they provide information on the accuracy of prior methodologies for predicting impacts. All of EA is based on the premise of predicting impacts from a facility that has not been built, or in the case of risk assessments of single chemicals, substances that have not entered the marketplace. Thus, monitoring actual releases into the environment, actual exposure paths and health outcomes is essential for understanding how good are the predictions. Poor predictions allow for adjustments to future methodologies involving similar types of facilities and/or substances. Nevertheless, feedbacks that suggest apparently accurate predictions should be interpreted cautiously, particularly when little time has passed.

For example, initial risk assessments of nuclear electricity generation which suggested risk of death was very low had to be adjusted to acknowledge the deaths and disability from the Chernobyl meltdown and more recently the one at Fukushima. Such data have also been kept for other major forms of electrical power generation, which helps produce data like that in Table 9.1, which still shows nuclear power to be lowest among the big four in terms of fatalities standardized by the number of years of power generation (TWy). These data are nevertheless very politically charged with considerable debate about the nuclear deaths in particular as described in Chapter 4. Further, Beranek (2011) writing for Greenpeace, criticizes nuclear energy–related death statistics the come from various OECD and nuclear associations reports (like the data in Table 9.1) for (i) being traced back to the work of only two authors with little cross-verification; (ii) often ignoring non–OECD countries (this allows omitting the Chernobyl disaster); and (iii) the

Table 9.1 (Disputed) fatalities due to electricity generation by source OECD and non–OECD countries combined

Energy Source	*Fatalities/Terawatt Year*
Coal[1]	1,797
Natural Gas[1]	196
Hydro[1]	10,288
Nuclear[1]	48*
Biomass[2]	2,400
Wind[2]	1.5
Solar[2]	4.4

[1]World Nuclear Association (2014)

[2]Bickel and Friedrich (2005), NRC (2010), Conca (2012)

*This figure in particular is highly disputed, highlighting that there is a complex politics of energy risk.

use of modelled deaths for the nuclear industry compared to actual deaths for others. He goes on to conclude that nuclear deaths are more likely on par with those of natural gas.

Debate aside, while the highly regulated nuclear industry, along with governments and academics, has kept an eye on the impacts of the electrical power industry over time, the same might not be true of other industries and developments. Unless an industry is compelled to report toxic emissions, they may prefer to keep data to themselves or save money and not collect it at all. Fortunately several countries like the U.S. and Canada have legislated mandatory public reporting of toxic releases that allows all stakeholders involved in disputes over risk to have access to emissions information (Environment Canada, 2014; EPA, 2014b). Such public repositories of information make it far easier for all stakeholders to have access to information, whereas in the past such data might have been kept within industry boundaries and only been made available through the EA process itself (Fung and O'Rourke, 2000). Yet there are dissenters as well who caution about the dangers of freely available toxic chemical information in a post-9/11 world (Dudley, 2004). Thus, environmental justice considerations, particularly the principle of 'right to know', can come into tension with national security concerns – underscoring further the complex politics of environmental health risk.

Alternatives to risk and environmental assessment

A key drawback to an adversarial EA system for assessing the impacts of projects on a local community is that it depends on a single, non-local judge or jury to deny or approve the undertaking. There are a number of alternatives to such a system that not only attempt to contain conflict but do so in the interest of coming to decisions that balance worldviews on how and whether to proceed with development. Although not all cases of EA necessarily involve disagreements on the best way forward when they do arise there are various forms of alternative dispute resolution (ADR) that may come into play. Examples of ADR include *environmental conflict resolution (mediation)* and *citizen juries*. In some cases these could theoretically be subsumed within existing hierarchical EA processes, but in other cases they would involve overhauling the system to more evenly share out decision-making power.

ADR is a suite of conflict resolution techniques that provides an alternative to the courts or court-like systems of decision making. It is generally associated with *deliberative democracy*, a wider philosophy of decision making. Deliberative democracy emphasizes the process of consensus building rather than merely voting – though voting, in some instances, may be part of the process. The EA process in Canada, for example, essentially functions like a form of *binding arbitration* in which all stakeholder parties to a decision empower a neutral third party to make the decision on their behalf. The stakeholders present evidence to the arbitrator – often a former judge or lawyer – and court-like procedures are used. However, in the case of EA, there is considerable emphasis on scientific (including social scientific) evidence of various sorts. EA is not a pure form of binding arbitration, because the arbitrator – usually an environment minister or their appointees – is generally not agreed to by the stakeholders; rather, they are appointed through existing legislative arrangements. That is, who *gets* to decide is pre-ordained long before the dispute arises. O'Leary et al. (2004) identify a number of characteristics that comprise ADR and which move away from this binding arbitration-style model:

i. participation in the process is voluntary;
ii. an independent, neutral third-party mediator must help the stakeholders come to an agreement;
iii. the mediator must *not* have the power to impose a decision;
iv. stakeholders (or their appointed experts) must be allowed to participate directly;
v. stakeholders must agree on the outcome through consensus;
vi. stakeholders may withdraw from the process and seek a decision through other legal means (e.g., the courts).

A cornerstone of ADR then is independent mediation. There is certainly a philosophical debate about whether independence is ever possible in a society in which, as we have already argued, values shape decisions. One way to address this philosophical quandary is to allow stakeholders to come to a consensus on who will be the mediator. What happens in practice is that this mutually agreed-upon mediator facilitates open dialogue between stakeholders, meeting with them together and as individual groups to help find common ground upon which a consensus can be built. This is particularly useful where large portions of the local community are cautiously open to discussion rather than vehemently and unconditionally opposed to a development.

Citizens' juries, on the other hand, share with existing EA that a decision is rendered rather than a consensus built. This is based largely on a pluralist philosophy of decision making that aligns with market-based economics. That is, in pluralist decision making, while all participants in a decision may be attempting to forward their own interests, participants find common ground where their interests intersect enough to make a decision path agreeable to all. Thus, the difference between citizen juries and traditional EA is that the decision panel is made up, at least in part, of local residents who will be directly impacted – both positively and negatively – by the development/undertaking. As with binding arbitration, the challenge of this form of environmental decision making is selecting the jury (Smith and Wales, 2000). Ideally citizen jurors would be randomly selected in some way, and under certain circumstances key stakeholders may be provided the opportunity to screen potential jurors. In contrast to environmental conflict resolution with mediation which comes to a consensus on the way forward, in the citizens' jury process the consensus is instead on who the decision makers will be. Unlike ECR and other forms of ADR, in which citizen participants essentially volunteer their time to be involved, in the case of citizens' juries they would be seconded from the ordinary daily activities through various legal mechanisms, which would likely include paying a salary. A major limitation of citizen juries is that they are prone to *groupthink* despite the ideal of pluralist decision making. Groupthink happens when people formally or subconsciously agree to think alike in order to minimize conflict. The problem is that particularly assertive people in the group can tilt decisions in their favour by making others so uncomfortable the remainder of the group simply agree to restore harmony. The tendency towards groupthink was one of the findings from an experimental citizen jury study conducted by Huitema et al. (2007) in the Netherlands. They found that observing the jury in action first and then subdividing the jurors into a talkative group and less talkative group helped even out the amount people ultimately spoke.

There are various ways these alternative techniques can be subsumed under existing hierarchical EA. For example, the province of Ontario in Canada promotes mediation to resolve conflicts within EA directed by the Ministry of Environment. Mediation can happen at any point in the process to resolve disputes between stakeholders, with an ideal result potentially being an overall decision that more closely resembles consensus. Nevertheless, this type of mediation is more pragmatically oriented towards nudging along stalled EA processes. That is, the mediation report merely feeds into the existing process, and the Minister (perhaps with the help of a panel) still is the one who is the final arbiter of decisions (e.g., Ontario MOE, 2014). There are many other ways the public can be involved in decision making; in fact Rowe and Frewer (2005) outline more than 100 of them (e.g., citizens' advisory committee, citizen review board, neighbourhood planning council, tele-voting or web-voting, priority setting committee, social audit). A key theme in many of these is that at the end of the day when conflict is central, full participation in decision making may be preferred to mere consultation and information sharing for a project decision over which locals otherwise have no control.

Conclusion

This chapter touches on a range of issues involved in the health risk assessment of individual chemicals and the environmental assessment of projects such as new facilities or already contaminated cleanup sites. There are two central themes: (i) social values are infused throughout both processes and (ii) there are roles to be played by the social sciences in understanding these complex decision processes. As with all forms of decision making the process that is ultimately used has to strike a balance between scarce resources that includes both money and time. For example, the current hierarchically based systems might be financially less costly than alternative systems, but the alternative systems, particularly those involving alternative dispute resolution, might have greater legitimacy. This is especially the case for locals who most directly experience the negative health impacts of environmental undertakings. The downside of ADR processes is that they generally take longer. Nevertheless, there is growing pressure to move away from hierarchical, conflict-oriented systems and towards forms of decision making that better empower local citizens using deliberative democracy–based consensus in various forms. This is one way to handle the growing concerns about environmental health justice discussed in Chapter 8. The next chapter discusses how current decision-making systems may lead to entrenching environment and health injustices and repeating past mistakes.

References

Agency for Toxic Substances and Disease Registry (ATSDR). 2015. *Appendix C: Chemical Mixtures Exposure.* Available at: http://www.atsdr.cdc.gov/HAC/pha/KellyAFB-PC101204/KellyAFB_appendC-D.pdf (Accessed 24 July 2015).

Ale, B.J.M. 2005. Tolerable or acceptable: A comparison of risk regulation in the United Kingdom and in the Netherlands. *Risk Analysis*, 25(2), 231–241.

Barnes, G., Baxter, J., Litva, A., & Staples, B. 2002. Beyond physical health effects: Interpersonal and community-level impacts of chronic chemical contamination. *Social Science and Medicine*, 55(12), 2227–2241.

Baxter, J., & Lee, D. 2004. Explaining the maintenance of low concern near a hazardous waste treatment facility. *Journal of Risk Research*, 6(1), 705–729.

Beranek, J. 2011. *Deaths and Energy Technologies, Greenpeace International.* Available at: http://www.greenpeace.org/international/en/news/Blogs/nuclear-reaction/deaths-and-energy-technologies/blog/34275/ (Accessed 24 July 2015).

Berger, T. 1977. *Northern Frontier, Northern Homeland–Mackenzie Valley Pipeline Inquiry Report–Volume 1 & 2.* Available at: https://docs.neb-one.gc.ca/ll-eng/llisapi.dll?func=ll&objId=238336&objAction=browse&redirect=3 (Accessed 20 October 2014).

Bickel, P., & Friedrich, R. 2005. *Externalities of Energy*, European Union Report EUR 21951, Luxembourg.

Canadian Environmental Assessment Agency (CEAA). 2014. *Cumulative Effects Assessment Practitioners' Guide.* Available at: https://www.ceaa-acee.gc.ca/default.asp?lang=En&n=43952694–1&offset=6&toc=show (Accessed 22 October 2014).

Carson, R. 2002. *Silent Spring*. Boston: Houghton Mifflin Harcourt.

Collins, J.J., & Lineker, G.A. 2004. A review and meta-analysis of formaldehyde exposure and leukemia. *Regulatory Toxicology and Pharmacology*, 40(2), 81–91.

Conca, J. 2012. How deadly is your kilowatt? We rank the killer energy sources. *Forbes Magazine Business*, June 10, 2012.

Conis, Elena. 2010. Debating the health effects of DDT: Thomas Jukes, Charles Wurster, and the fate of an environmental pollutant. *Public Health Reports*, 125(2), 337.

Consonni, D., Pesatori, A.C., Zocchetti, C., Sindaco, R., D'Oro, L.C., Rubagotti, M., & Bertazzi, P.A. 2008. Mortality in a population exposed to dioxin after the Seveso, Italy, accident in 1976: 25 years of follow-up. *American Journal of Epidemiology*, 167(7), 847–858.

Cooper, L.M., & Sheate, W.R. 2002. Cumulative effects assessment: A review of UK environmental impact statements. *Environmental Impact Assessment Review*, 22(4), 415–439.

Cormier, S.A., Lomnicki, S., Backes, W., & Dellinger, B. 2006. Origin and health impacts of emissions of toxic by-products and fine particles from combustion and thermal treatment of hazardous wastes and materials. *Environmental Health Perspectives*, 114(6), 810–817.

Crighton, E.J., Elliott, S.J., Van Der Meer, J., Small, I., & Upshur, R. 2003. Impacts of an environmental disaster on psychosocial health and well-being in Karakalpakstan. *Social Science & Medicine*, 56(3), 551–567.

Deacon, L., & Baxter, J. 2013. No opportunity to say no: A case study of procedural environmental injustice in Canada. *Journal of Environmental Planning and Management*, 56(5), 607–623.

Dudley, S.E. 2004. Is it time to reevaluate the toxic release inventory. *Missouri Environmental Law and Policy Review*, 12, 1.

Durant, R.F., Fiorino, D.J., & O'Leary, R. (Eds.). 2004. *Environmental Governance Reconsidered: Challenges, Choices, and Opportunities*. Boston: MIT Press.

Eckerman, I. 2005. *The Bhopal Saga: Causes and Consequences of the World's Largest Industrial Disaster*. Hyderabad, India: Universities Press.

Edwards, J.G. 2004. DDT: A case study in scientific fraud. *Journal of American Physicians and Surgeons*, 9, 83–88.

Environment Canada. 2014. *National Pollutant Release Inventory.* Available at: https://www.ec.gc.ca/inrp-npri/ (Accessed 20 October 2014).

Environmental Protection Agency (EPA). 2014a. *Risk Assessment.* Available at: http://www.epa.gov/risk_assessment/index.htm (Accessed 6 October 2014).

Environmental Protection Agency (EPA). 2014b. *The Toxics Release Inventory (TRI) Program.* Available at: http://www2.epa.gov/toxics-release-inventory-tri-program (Accessed 6 October 2014).

Eyles, J., & Fried, J. 2011. Breaking the connections: Reducing and removing environmental health risk in the Canadian nuclear power industry. *Environmental Health and Biomedicine*, 15, 59.

Fischer, T.B. 2003. Strategic environmental assessment in post-modern times. *Environmental Impact Assessment Review*, 23(2), 155–170.

Foster, K.R., Vecchia, P., & Repacholi, M.H. 2000. Science and the precautionary principle. *Science*, 288(5468), 979–981.

Fung, A., & O'Rourke, D. 2000. Reinventing environmental regulation from the grassroots up: Explaining and expanding the success of the toxics release inventory. *Environmental Management*, 25(2), 115–127.

Hansen, J., Holm, L., Frewer, L., Robinson, P., & Sandøe, P. 2003. Beyond the knowledge deficit: Recent research into lay and expert attitudes to food risks. *Appetite*, 41(2), 111–121.

Health Canada. 2014a. *A Framework for Strategic Risk Communications Within the Context of Health Canada and the Public Health Agency of Canada's Integrated Risk Management*, Available at: http://www.hc-sc.gc.ca/ahc-asc/pubs/_ris-comm/framework-cadre/index-eng.php (Accessed 6 October 2014).

Health Canada. 2014b. *Government of Canada Radon Guidelines*. Available at: http://www.hc-sc.gc.ca/ewh-semt/radiation/radon/guidelines_lignes_directrice-eng.php (Accessed 20 October 2014).

Huitema, D., van de Kerkhof, M., & Pesch, U. 2007. The nature of the beast: Are citizens' juries deliberative or pluralist? *Policy Sciences*, 40(4), 287–311.

Jasanoff, S. 2005. *Designs on Nature: Science and Democracy in Europe and the United States*. Princeton, NJ: Princeton University Press.

Jeffery, M.I. 2002. Intervenor funding as the key to effective citizen participation in environmental decision-making: Putting the people back into the picture. *Arizona Journal of International & Comparative Law*, 19, 643.

Johnson, F.R., & Smith, V.K. 2013. 'Informed Choice or Regulated Risk? Lessons from a Study in Radon Risk' in Glickman, T. & Gough, M. (eds.). *Readings in Risk*. Washington DC: Resources for the Future. p247.

Kasperson, R.E., Renn, O., Slovic, P., Brown, H.S., Emel, J., Goble, R., . . . & Ratick, S. 1988. The social amplification of risk: A conceptual framework. *Risk Analysis*, 8(2), 177–187.

Kortenkamp, A. 2007. Ten years of mixing cocktails: A review of combination effects of endocrine-disrupting chemicals. *Environmental Health Perspectives*, 115(Suppl 1), 98–105.

Lazarus, R.S., & Folkman, S. 1984. *Stress, Appraisal, and Coping*. New York: Springer.

Leiss, W. 1996. Three phases in the evolution of risk communication practice. *The Annals of the American Academy of Political and Social Science*, 545, 85–94.

Lerche, I., & Glaesser, W. 2006. *Environmental Risk Assessment*. Berlin: Springer.

Lerer, L.B. 1999. Health impact assessment. *Health Policy and Planning*, 14(2), 198–203.

Létourneau, E.G., Krewski, D., Zielinski, J.M., & McGregor, R.G. 1992. Cost effectiveness of radon mitigation in Canada. *Radiation Protection Dosimetry*, 45(1–4), 593–598.

Lewis, R.E., & Tyshenko, M.G. 2009. The impact of social amplification and attenuation of risk and the public reaction to mad cow disease in Canada. *Risk Analysis*, 29(5), 714–728.

López-Carrillo, L., Torres-Arreola, L., Torres-Sánchez, L., Espinosa-Torres, F., Jiménez, C., Cebrián, M., . . . & Saldate, O. 1996. Is DDT use a public health problem in Mexico? *Environmental Health Perspectives*, 104(6), 584.

Moeller, D.W., 2011. *Environmental Health*. Boston: Harvard University Press.

National Research Council (US)., Committee on Health, Environmental, and Other External Costs and Benefits of Energy Production and Consumption. 2010. *Hidden Costs of Energy: Unpriced Consequences of Energy Production and Use*. Washington, DC: National Academies Press.

O'Brien, M. 2000. *Making Better Environmental Decisions: An Alternative to Risk Assessment*. Cambridge, MA: MIT Press.

O'Leary, R., Nabatchi, T., & Bingham, L. 2004. 'Environmental Conflict Resolution (Ch 9).' in Durant, R.F., Fiorino, D.J., & O'Leary, R. (eds.). *Environmental Governance Reconsidered: Challenges, Choices, and Opportunities*. Boston: MIT Press. pp 323–354.

Ontario Ministry of Environment. 2014. *Using Mediation in the Environmental Assessment Process*. Available at: https://www.ontario.ca/environment-and-energy/using-mediation-ontarios-environmental-assessment-process (Accessed 9 October 2014).

O'Shaughnessy, P.T. 2008. Parachuting cats and crushed eggs the controversy over the use of DDT to control malaria. *American Journal of Public Health*, 98(11), 1940.

Ozonoff, D. 1994. Conceptions and misconceptions about human health impact analysis. *Environmental Impact Assessment Review*, 14(5), 499–515.

Pesticide Management Regulatory Agency. 2014. Available at: http://www.hc-sc.gc.ca/ahc-asc/branch-dirgen/pmra-arla/index-eng.php (Accessed on June 11, 2014).

Pidgeon, N., Kasperson, R.E., & Slovic, P. (Eds.). 2003. *The Social Amplification of Risk*. New York: Cambridge University Press.

Poortinga, W., Bronstering, K., & Lannon, S. 2011. Awareness and perceptions of the risks of exposure to indoor radon: A population-based approach to evaluate a radon awareness and testing campaign in England and Wales. *Risk Analysis*, 31(11), 1800–1812.

Poortinga, W., & Pidgeon, N.F. 2004. Trust, the asymmetry principle, and the role of prior beliefs. *Risk Analysis*, 24(6), 1475–1486.

Raffensperger, C., & Tickner, J.A. (Eds.). 1999. *Protecting Public Health and the Environment: Implementing the Precautionary Principle*. Washington, DC: Island Press.

Renn, O., & Rohrmann, B. 2000. *Cross-Cultural Risk Perception*. Boston: Kluwer Academic.

Rowe, G., & Frewer, L.J. 2005. A typology of public engagement mechanisms. *Science, Technology, & Human Values*, 30(2), 251–290.

Rushton, L. 2003. Health hazards and waste management. *British Medical Bulletin*, 68(1), 183–197.

Sabatier, P.A., & Jenkins-Smith, H.C. (Eds.). 1993. *Policy Change and Learning: An Advocacy Coalition Approach*. Boulder, CO: Westview Press.

Schwarz, M., & Thompson, M. 1990. *Divided We Stand: Re-Defining Politics, Technology and Social Choice*. Philadelphia: University of Pennsylvania Press.

Simon, T. 2014. *Environmental Risk Assessment: A Toxicological Approach*. Boca Raton, FL: CRC Press.

Slovic, P. 1993. Perceived risk, trust, and democracy. *Risk Analysis*, 13(6), 675–682.

Smith, G., & Wales, C. 2000. Citizens' juries and deliberative democracy. *Political Studies*, 48(1), 51–65.

Stern, P.C., & Dietz, T. (Eds.). 2008. *Public Participation in Environmental Assessment and Decision Making*. Washington, DC: National Academies Press.

Stockholm Convention. 2014. *Convention Text*. Available at: http://chm.pops.int/TheConvention/Overview/TextoftheConvention/tabid/2232/Default.aspx (Accessed 10 October 2014).

Teräväinen, T., Lehtonen, M., & Martiskainen, M. 2011. Climate change, energy security, and risk–debating nuclear new build in Finland, France and the UK. *Energy Policy*, 39(6), 3434–3442.

Therivel, R., & Ross, B. 2007. Cumulative effects assessment: Does scale matter? *Environmental Impact Assessment Review*, 27(5), 365–385.

Wiener, J.B., & Rogers, M.D. 2002. Comparing precaution in the United States and Europe. *Journal of Risk Research*, 5(4), 317–349.

WHO. 2011. *The Use of DDT in Malaria Vector Control: WHO Position Statement*. Available at: http://apps.who.int/iris/handle/10665/69945 (Accessed 9 October 2014).

World Nuclear Association. 2014. *Safety of Nuclear Power Reactors*. Available at: http://www.world-nuclear.org/info/Safety-and-Security/Safety-of-Plants/Safety-of-Nuclear-Power-Reactors/ (Accessed 20 October 2014).

Wynne, B. 2006. Public engagement as a means of restoring public trust in science–hitting the notes, but missing the music? *Public Health Genomics*, 9(3), 211–220.

10 And more of the same?

Introduction

Are we doomed to repeat environment and health disasters; to repeat 'failures' from the past? The national headlines in Canada at the time of writing include the tragic exposure of five children to the pesticide phosphine in Fort McMurray, Alberta, with two children succumbing to the toxic inhalation exposure (CBC, 2015a). This comes in the form of tablets used for fumigation for bedbugs in some countries – particularly tropical places like Pakistan and Thailand where bedbugs are a serious health problem – but they are banned for such use in Canada. The first child that died was the youngest of all of the children, and therefore she was likely the most vulnerable. This may be a case of failsafe systems failing (see Chapter 5), but what is to prevent this tragedy from being repeated with this or other chemicals? Coincidentally the same toxin and scenario have been implicated in the deaths of two sisters travelling in Thailand according to one coroner, who affirms the role of uncertainty by saying, "Science doesn't allow us to confirm this without a doubt. Maybe one day it will" (CBC, 2015b). But as we have shown, uncertainty of some kind will always persist.

It seems we will continue to produce harmful chemicals to be approved for use under certain conditions when the benefits are purportedly high. At the same time we may tacitly or overtly blame the victim for illegal use or for not using highly toxic chemicals as directed. In a study of cosmetic pesticide bans in Canada, for example, Hirsch and Baxter (2009) learned that despite a by-law prohibiting the use of a long list of pesticides in Halifax, some residents continued to use them and did not use the pesticides as approved by the national health agency. Pesticides regulation itself is symbolic of the way social values are heavily inscribed on the science (e.g., toxicology and epidemiology) whereby there is a patchwork of allowed and disallowed substances and uses on multiple scales from nation states right down to the municipal level. As we have seen throughout the examples in this book the same is true for policies relating to such things as greenhouse gases, methylmercury, arsenic, asbestos and nuclear power. What social science can add is better understanding of the social determinants of behaviours that impact both individual and collective attitudes and actions towards these environmental substances and phenomena that impact health.

One way forward in our analysis of environment and health is to consider two pathways into the future. The first is a path of repeating our past missteps despite knowledge of harm. The second is a path of learning from our missteps to minimize future harms. The latter is clearly a more optimistic scenario, and in the spirit of tying things together, we will briefly explore how these align with concepts and examples already described throughout this book. Before tracing these potential paths forward, the stage is set in the next section, which more thoroughly outlines a model of environmental health impacts by moving beyond physical, somatic symptoms and their determinants.

Only physical symptoms matter?

A theme in this book that creates space for the social sciences is a broad view of health. Though the World Health Organization definition of health as not just physical well-being but also mental and social well-being has ostensibly been embraced worldwide and across academic disciplines, there seem to be implicit and sometimes explicit assumptions that physical health is all that *really* matters in risk and environmental assessment. Figure 10.1 outlines a general conceptual model of environment and health that incorporates such a definition. It summarizes a heuristic for thinking about health impacts from a facility, but it could be adapted to all manner of environmental health threats. What the diagram highlights is that direct somatic effects from what we typically consider to be the exposure – for example, chemicals, radiation – tells only part of the story. The impacts are multiple and are mediated in some way by what we have been calling 'social' factors, whereby for example decisions about facility siting or routes of freighter traffic as well as felt local benefits may have a strong bearing on how things like chemical and aesthetic exposures are interpreted. In this sense the facility siting process may actually comprise part of the 'exposure'. For example, local residents may feel that the siting process for a facility proposed for their community is unfair (e.g., other sites not considered) or unjust (e.g., the community is socially disadvantaged). Similarly, the effects of these exposures may be mediated by other determinants. For example, various types of benefits – financial, social – may offset impacts of exposure, while at the same time certain sensitivities – noise, odours – may increase the effects of these exposures. Along with benefits, the siting process itself can have a strong influence on felt impacts as in the example of the 'successful'[1] voluntary, open and largely democratic high-level nuclear waste disposal siting efforts in Sweden (Sjöberg, 2004). Allowing municipalities the freedom to run their own local discussion and referenda on willingness to host allowed an exploration of local values on the issue. For those municipalities that ultimately 'competed' for the facility, national responsibility and even pride became a prominent theme motivating ongoing inclination to be wards of nuclear waste (Sundqvist, 2002).

At the bottom of the figure there are two types of tightly interconnected outcomes – what we are calling Responses (e.g., annoyance, stress) and Impacts (e.g., disease, well-being impacts). This is where the WHO definition of health is incorporated. Thus, annoyance, while it may not be considered by some policy makers and facility-siting agents to be a health impact, because it is possibly connected to well-being outcomes including disease, it is certainly something to consider for mitigation (see also Shepherd et al., 2011). It is currently unclear to what degree such an expanded model of environment and health impacts is considered operationally in research and decision making about environmental hazards.

If the impacts are mainly psychosocial there seems to be a tendency to downgrade their importance. As stress, anxiety, frustration and social conflict are normal aspects of everyday life, does the amount of these added to a community from disaster or facility siting matter enough to warrant mitigation? This has not so far been the case concerning the health impacts of wind turbines, as annoyance tends to be considered less important than physical symptoms by siting agents, developers and turbine operators. The backlash for denigrating concerns about health (and fair siting) of turbines has been extensive, as in Ontario, where almost a quarter of the municipalities have passed council votes saying they are unwilling to host turbines (Walker et al., 2014). However, as the Swedish case of the high-level nuclear facility demonstrates, there may yet be win-win scenarios available to wind and other technology developers who face head on and mitigate impacts such as annoyance and the values that undergird them.

The health impacts of wind turbines are an interesting case since this technology is billed as one solution to greenhouse gas emissions. The growing literature includes implicit and explicit discussion of the *nocebo* effect as an explanation for high amounts of symptom reporting around turbines (e.g., headaches, dizziness). A nocebo effect happens when an inert substance causes an adverse health outcome, the same way a placebo may cause a positive health outcome. The usual explanation is that the mechanism for the negative effect is psychosomatic. For example, a study of a community with high symptom reporting near a cell phone tower that had never been turned on has had considerable mileage in ostensibly demonstrating this effect in facilities siting (Bansal, 2009). There are two ways to look at this. The first is to say that the risks in such cases are not 'real' (Page et al., 2006), the implication being that nothing further needs to be done in terms of prevention and mitigation. This seems to be the stance that Chapman and colleagues are taking in relation to wind turbines (Chapman et al., 2013; Crichton et al., 2014). A second interpretation is to ask further questions about the broader context that may be contributing to the negative impact reporting. In the case of wind turbines Chapman et al. suggest that it is the hype and vitriol surrounding the health impacts of turbines that either suspiciously or ironically results in higher symptom reporting. Overall there are questions about what exactly is the exposure, and facility siting is a good candidate because it, among other things, is considered problematic in several jurisdictions (Devine-Wright, 2005). Regardless, the different interpretations of nocebo effects underscores that environment and health will remain highly politicized. The next sections return to the issue of possible paths forward in the context of such a model of environmental and health impacts.

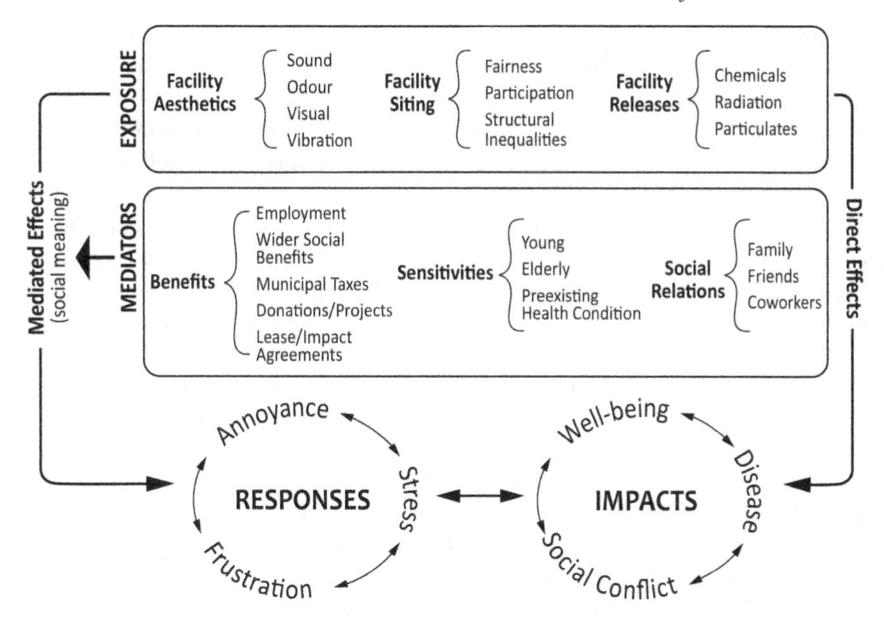

Figure 10.1 Direct and mediated social pathways for environment and health impacts
Source: Adapted from Shepherd et al. (2011).

Path 1: Repeating tragedies and cognitive dissonance

The first path, repeating past mistakes, is of course the most troubling, as the case of phosphine poisoning exemplifies. This suggests that we simply have to get used to such stories, as accidents are a 'normal' and unfortunate by-product of industrial capitalism. However, such problems are also repeated on a global scale where even more lives are at stake. For example, the meltdowns of Three Mile Island and Chernobyl did not teach enough lessons to prevent the disaster at Fukushima. No doubt many people feel less safe in the wake of these disasters, while others may feel nothing much has changed. Are the three nuclear plants which succumbed to disaster so different from other nuclear plants? Can it happen here (wherever that may be) too? What helps people avoid the uncomfortable feeling that it could happen at their own local nuclear power facility typically comprises making a list of all the differences between the context of the disaster facilities and the local facility – for example, non-coastal, no local fault lines, different reactor design and different failsafe systems. This is the emotion-focused coping described in Chapter 6, a process that helps us avoid the uncomfortable feeling that things may be more dangerous than we imagine. Meanwhile the nuclear industry in places like Canada is slowly admitting that disasters are likely, so that they are switching their safety maximizing efforts as follows: "the emphasis in the nuclear industry was on design and prevention, but now it's on prevention and mitigation" (Bissett, 2015). Among other things this means diesel backup power systems on site to run emergency cooling procedures.

While the nuclear industry admits they now "expect the unexpected" (Bissett, 2015), there are psychosocial processes at work that may prevent such an admission in the general public. This is a phenomenon known in the social sciences, particularly psychology, as cognitive dissonance – the disorienting and uncomfortable feeling when new information conflicts with deeply held beliefs and values (Festinger, 1957). If we accept the premises of these two theories then, we are doomed to repeat environment and health disasters if people favour emotion-focused coping meant to reduce dissonance rather than taking action (e.g., to phase out nuclear technology altogether). That is, if people simply mentally re-sort phenomena to fit existing core beliefs and values that support the status quo we are less motivated as a society to move away from too-dangerous technologies and look for less harmful alternatives. Of course 'too dangerous' itself is socially defined, linked to the same sets of core values that lead to emotion-focused coping and dissonance-avoiding mental calculus in the first place.

Cost-benefit analysis when combined with risk analysis has been portrayed as a rational way to make decisions about environmental health hazards, but there is concern that this approach does not adequately account for a range of costs (Chapter 4). Overall, the costs of taking on a decision path that allows ongoing exposure – whether that be maintaining the status quo (e.g., bisphenol A, thalidomide, existing nuclear energy facilities) or introducing a new product or technology (e.g., GMOs, oil pipelines, new nuclear facilities) – tends to over-emphasize the benefits and underplay the costs. Costs may inappropriately be measured in terms of mortality alone, which leaves out all of the chronic outcomes that can negatively impact well-being. Beyond the distasteful idea of putting a dollar value on human life cost-benefit analysis discounts costs like chronic conditions, loss of way of life as well as psychosocial well-being (Ackerman and Heinzerling, 2002).

There is also concern that the marginalized in society are destined to always bear a disproportionate burden of environment and health harms as a result of rational procedures like cost-benefit analysis. In this volume we trace a tension between Beck's risk society (Chapter 5) idea that 'nobody escapes' globalized hazards and the growing concern about environmental justice – that industrial harms are meted out disproportionately on the poor, racial and other marginalized groups in society (Chapter 8). The Bhopal disaster which killed thousands and injured tens of thousands from extremely toxic methyl isocyanate (MIC) exposure near Union Carbide's pesticide plant in India is considered to be one of the worst industrial disasters in history, yet pesticide plant disasters continue. For example, in the small, relatively poor historically farming town of West in Texas, the West Fertilizer Co. plant exploded in 2013, killing 15 and wounding 200+ others under conditions deemed 'preventable' by a safety board (McLaughlin, 2013). Twelve years earlier, but decades after Bhopal, a similar explosion killed 31 locals and injured 2,000+ (Matlack, 2013). Though the latter two examples were not due to leakage of MIC, but instead the volatile aspects of the ammonium nitrate, they all have in common the failure of complex systems. They also have in common that they disproportionately impacted poorer communities and, in the case of Bhopal, a relatively poor country. Beck suggests we are moving away from emotion-focused coping, as people are increasingly paying attention to and railing against substances, facilities and technologies they feel threaten health. Yet this resistance comes from a wide array of communities, not just marginalized groups.

There are very different approaches to bolstering safety through policy. Sticking with the fertilizer risk example, while the U.S. has focused their policy efforts on the sale and storage of ammonium nitrate, largely due to the Oklahoma City bombing that killed 168, France has focused their efforts on creating buffers around facilities handling this chemical to minimize human impacts around facilities if and when an explosion/fire does happen. Meanwhile in India, the Bhopal MIC disaster motivated their first Environmental Protection Act (1986), but regulations controlling multinational industries have been applied in such a manner as to keep foreign capital flowing to the region to stimulate a rapidly growing economy. Though disaster preparedness has improved in places like Bhopal, this is in the context of communities that have other competing and basic public health problems to manage like public drinking water, sewage and waste disposal (Broughton, 2005). This is set against a post-disaster backdrop where though the deaths from the disaster are undisputed, sorting out the long-term health impacts is problematic. There are constraints in being able to use the strongest study designs due to bias from widespread coverage of the disaster (e.g., using self-assessed health measures) and problems with exposure classification amid numerous potential confounders in this relatively poor community (Dhara and Dhara, 2002).

This case of repeated fertilizer plant disasters suggests that fail-safe systems that depend on the use of substances 'as directed' are only part of the story of human health and chemical safety. There is much to be learned about the social determinants of actions and policies that support social and technological configurations that allow such risks to be posed in the first place.

Industrial disasters are not exclusive to the pesticide chemicals industry, but pesticide disasters have the dubious distinction of causing acute, tragic and hence readily identifiable short-term health outcomes. For other disasters the human health impacts may be more diffuse, longer term and even more difficult to trace. The BP *Deepwater Horizon* oil platform explosion and spill in the Gulf of Mexico that started in April 2010 is currently the worst ocean oil spill disaster in history, killing 11 in the explosion and releasing 4.9 million barrels of oil that devastated tourism, fishing and other coastal industries in the U.S., particularly in Louisiana. The financial impacts were enormous for the impacted communities and BP alike (at least $42 billion), but the wider spread of health impacts of oil spills are not always as direct, as in the case of an acute explosion or acutely toxic chemical (Fontevecchia, 2013).

For example, though Lyons et al. (1999) find statistically significant health impacts of the *Sea Empress* oil tanker disaster on the southern coastal communities of Pembrokeshire in Wales the impacts measured were a combination of mental (anxiety, depression) and physical (headaches, sore eyes and throat). Further, the statistical modelling controlled for the *mental* health impacts in order to determine that they did not alone cause the physical symptoms. This suggested that more was going on than a psychosomatic response. The physical symptoms happened both in the presence of mental health problems and in their absence (Lyons et al., 1999), indicating the need to address both separately. Physical health problems can be psychosocially mediated by the trauma of disaster – a common tack for suggesting that the symptoms will subside once the disaster passes – but they may also happen directly, indicating the need for very different forms of remediation to address toxic exposures (see Figure 10.1). Regardless, the overall impact of these disasters extends beyond acute loss of life and dollars lost as collected in databases like EMdat. For example that database documents the impacts of the *Deepwater Horizon* disasters as 11 deaths, economic losses of $2,000,000 and the total number of people affected at only 17 (CRED, 2015)! Such databases tend to focus on the acute, short-term effects of, in this case, the explosion but do not account for the wider spread of impacts over time and space. Similarly, in a major World Health Organization multi-expert report on the health impacts of Chernobyl the introductory section on page 2 begins with an account of the 30 *acute* deaths known to be caused by the disaster. Though this is followed by a description of thousands of unexpected thyroid cancers the fact that people are not showing up to work is dismissed as unlikely the direct result of radiation-caused disease:

> Some investigators have interpreted a temporary loss of ability to work among individuals living in contaminated areas as an increase in general morbidity. High levels of chronic disease of the digestive, neurological, skeletal, muscular and circulatory systems have been reported. However, most investigators relate these observations to changes in the age structure, the worsening quality of life, and post-accident countermeasures such as relocation.
>
> (WHO, 2005, p. 4)

Thus, much of the work showing significant morbidity effects is dismissed because the studies are based on weak ecological studies rather than on individual radiation dosimetry. Work based on self-reported health impacts with a focus on psychiatric disorders suggests that the impacts are both mental and physical and that teasing out the separate effects is exceedingly difficult (Havenaar et al., 1997). The tendency to focus on what is *most* 'knowable' and 'confirmable' – as acute mortality – will remain a problem for impact reporting and therefore policy learning.

Perhaps most disturbing about the *Deepwater Horizon* and *Sea Empress* cases, as with the pesticides industry disasters, are suggestions that market conditions are such that there is a real risk that other disasters on the same scale are likely to recur ("Obama oil spill commission's final report," 2011). Such disasters, though, are emboldening voices speaking out against the use of big oil projects to satisfy energy needs. New large-scale projects are being thoroughly scrutinized and politicized, with the recent Keystone Pipeline proposal receiving a presidential veto from U.S. president Obama because it encourages the energy-intensive and polluting processes required for extraction in the Alberta oilsands ("Keystone XL and the president's veto," 2015).

The 'smoking gun' cases like those described tend to grab the headlines as far as environment and health disasters are concerned, but there are numerous cases of much slower disasters, like the Aral Sea (Chapter 2) and methylmercury (Chapter 3). Like Chernobyl, in both of those cases the problem is more insidious, measured in years not hours or days from an acute event. In the case of methylmercury, one of the reasons for delayed regulation was an overdependence on certain forms of evidence for decision making. That is, the lack of undisputable evidence of harm in a lab or in acute epidemiologic studies was used as reason enough to maintain the status quo, do nothing and not implement controls on the substance. Though scientific conservatism serves science very well, it does not necessarily serve safety-assuring environmental policy well since it is by the nature of its basic premises disinclined to accept that there is an effect. Though regulatory agencies have acknowledged this in more recent years, the fact that it took 12 decades to control methylmercury should be instructive and perhaps worrying. The problem is that we may be asking too much of science vis-à-vis policy which leads to social suspicion of the latest 'epidemiologic scare', no matter how good the evidence. To muddy the waters even further the same events and substances may be interpreted differently depending on context. Such is the case with DDT, which still remains a useful tool against malaria-carrying mosquitos in equatorial regions but is banned for use where mosquitos are vectors for relatively less devastating diseases. In this sense, current WHO policies supporting the use of DDT in equatorial regions can make sense alongside a Canadian and U.S. ban supported by the likes of Rachel Carson's book *Silent Spring*. Her landmark book suggested that products like DDT had been indiscriminately and harmfully overused in North America for years (see Chapters 5, 9). Nevertheless, such differences have allowed DDT to remain a flashpoint for controversy, with some maintaining it never should have been banned and that it is emblematic of the limitations of the precautionary approach to policy.

Path 2: Learning from missteps

Precaution and the spectre of false negatives

The precautionary approach to policy is implied throughout this book as an approach that seems under-utilized. Yet, there are increasing instances where this approach is taken in a relatively timely manner – for example, bisphenol A. These are nevertheless, decisions made under considerable scientific uncertainties – the ones described in Chapters 1 through 4. Amidst various social pressures (e.g., purported benefits of a substance/technology) and economic pressures (e.g., large investments in research and development), it may take considerable courage to say no to a substance or technology, particularly one already in wide use. The book highlights as well that instances where experts invoke precaution in the face of scientific uncertainty are context dependent where, for example, the risks and benefits are shared by the same group (e.g., new drugs). The problem for environment and health risks is that more often there is a spatial and temporal scalar disconnect in the sense that benefits are often diffuse, while the risks are localized to particular times/places where substances are produced (e.g., pesticides facilities) and technologies deployed (e.g., nuclear plants). This partially explains the widespread success of the Montreal Protocol on ozone layer–destroying CFCs and the mixed success of the Kyoto and subsequent Protocols for reducing greenhouse gas emissions (Chapter 5). Ease of substitution was among the main reasons for the success in replacing CFCs, and the relative lack thereof has been responsible, at least in the short term, for slow replacement of high–greenhouse gas–emitting (energy) technologies. Further, there will likely never be 'enough' science to categorically say some substances/technologies are safe/unsafe. For example, the case of electromagnetic fields and health described further on in this chapter has been highly studied and policy remains mixed and somewhat tentative.

The case of thalidomide is telling about the potential pitfalls of assuming some science indicating *no* harm is sufficient to allow controlled use of a substance. This is not an environment and health case, but the story is instructive vis-à-vis the use of toxicological evidence in particular. Thalidomide is a drug still in use today for the beneficial treatment of certain forms of cancer, but it turned out to be the culprit in a generation's worth of birth deformities in numerous countries including Canada and much of Europe but, most notably, *not* in the U.S. The drug was prescribed for use by expectant mothers to control the debilitating effects of morning sickness and was considered otherwise harmless because the drug makers "could not find a dose high enough to kill a rat" in their LD50 studies (see Chapter 1; Fintel et al., 2009). Human trials at the time consisted of distributing the drug through physicians and then asking those same physicians to track their patients for adverse effects – rather than the current drug trials system, which is under much tighter control. The U.S. did not approve the drug because Dr. Frances Kelsey, a physician and pharmacologist working for the FDA at the time and the one responsible for the thalidomide file, was concerned that existing research did not test if the drug passed through the placental barrier (Rehman et al., 2011). This movement of the drug to the foetus indeed was a root cause of the short limbs and other birth deformities from thalidomide.

Two other aspects of this case are notable in terms of learning and action. First, within about a year of doctors prescribing the drug it was pulled for use by pregnant women because physicians readily linked thalidomide with mounting numbers of deformed babies. As with the case of bisphenol A in which baby formula containers were suspect, because children were involved action was relatively swift and precautionary. There seems to be more resistance to change when substances are not so selective. Second, the disaster motivated a more stringent drug trials system under the auspices of the Food and Drug Administration (FDA) and changed thinking about more thorough consideration of sensitive groups rather than the average adult (Rehman et al., 2011). In this sense risk analysts should be relying less on LD50 trials to find doses that kill lab animals and focus on a wider range of plausible biological mechanisms for harm. The problem of course is that the mechanisms are many and resources for testing are limited.

Precaution stifles innovation?

There may be concern that taking a precautionary approach in all cases of uncertainty will stifle technological innovation because the financial risks of bringing a product to market, for example, will be too uncertain. Further, Todt and Lujan (2014) explore the notion that the precautionary principle itself is anti-science in the sense that it does not respect scientific results that do *not* find causal associations. However, they conclude that in most instances the precautionary principle is consonant with science in the sense of being conservative in terms of decision making. Rather than assume something is safe though because a toxicological experiment has yet to be run to show it is not, the precautionary advocates feel the assumption should instead be that it is not safe and, nevertheless, to keep looking. In this sense the precautionary principle encourages high-quality repeated study of the substance to falsify the *not* safe null hypothesis; and proceed to produce a preponderance of evidence that such a hypothesis should (tentatively) be rejected before a product goes to market. Such an approach overtly challenges the idea of failsafe itself and instead implies the need for more and more science – for example, toxicology, epidemiology and allied exposure sciences (e.g., atmospheric sciences, hydrologic sciences) – to *increase* confidence in safety. This is the rationale for Canada's Pesticide Management Review Agency re-assessing data on a wide range of chemicals already on the market, and it also plays a role in the story that follows of the health impacts of electromagnetic fields (see also Foster et al., 2000). The space shuttle *Columbia* disaster (2003) happening despite efforts to learn lessons from the *Challenger* disaster (1986) remind us that even the keenest attempts at safety from top scientists cannot prevent harm, and this is not necessarily due merely to chance (unknown unknowns; Vaughn, 2009).

Precaution for whom?

Though new technologies may be framed as opportunities, this is in tension with concerns about environmental justice, especially when they are accompanied by a hazardous facility like the fertilizer plant in Bhopal. For example, though poor and otherwise marginalized communities may be exposed to the negative externalities of facilities they may also benefit from those same facilities, for example, in terms of employment. Alhakami and Slovic (1994) point out that this idea extends to the technology itself, such that perceived benefit confounds risk perception. If individuals perceive a technology to be beneficial to society, they are less apt to judge it as a risk in need of further regulation simply *because* of the benefit (e.g., medicines, electricity, cell phones).

Swan Hills, Alberta, the host of a hazardous waste incinerator, is a case in point. The facility was put in place using a voluntary siting process in which communities competed for the facility amid the promise of an improved tax base, stable jobs and overall community improvements. All of the communities at the time were experiencing the effects of economic downturn and loss of population, and 120 steady jobs was appealing in the volunteering towns. In a community vote the residents of Swan Hills, a rural, largely resource-based town dependent on forestry and oil and gas extraction, was largely in favour of hosting the facility, and that support has remained over time despite a fire, chemical leaks, hunting/fishing bans due to leaks and elevated PCB in the blood of workers (Baxter and Lee, 2004). Though the accidents at the facility garnered provincial and national headlines and the highest ever environmental fine in the province at the time, there has been little evidence of adverse health impacts in humans. Nevertheless, no peer-reviewed epidemiologic studies of the community and the facility have been published. Of course one of the challenges in any such case is that the population is too small (about 2,000) for detecting relatively rare health impacts like cancers or birth defects, so the risk of false negatives may deter such a study in the first place.

The second path – learning from our missteps – aligns well with Sabatier's optimistic notions of policy learning through mediated knowledge. That is, he suggests better policy outcomes happen when divergent coalitions of stakeholders – one typically advocating for further restrictions and a precautionary approach to policy and the other championing new technologies or developments with minimal restrictions – interact. As suggested by the cultural theory of risk, the former group tend to be oriented to egalitarian thinking and the idea that the environment and humans tend to be more fragile than current policy recognizes, while the latter group are more apt to share an individualist worldview characterized by entrepreneurial risk taking and the assumption that nature is generally elastic and can bounce back from various types of environmental harms. In the Alberta hazardous waste facility example, residents in the town of Swan Hills tended toward the latter view while First Nations and surrounding communities tended toward the former. Yet in the process for siting at Swan Hills one of the safety decisions was to locate the toxics incineration facility 15 km away from any town (including Swan Hills). Environmental assessment – a relatively recent phenomenon in the history of industrialization – can thus work in favour of Sabatier's policy learning idea if it allows for this type of compromise. However, those towns that are much farther away, some downwind of the facility, that were not allowed a say in the community vote, remain the most concerned about all manner of environmental impacts including human health (Baxter and Greenlaw, 2005). At least one of these communities is made up of First Nations, and though they have received funds to support such things as scholarships in the community, concern about the facility remains high, particularly in view of hunting and fishing bans. This suggests that though voluntary siting combined with public environmental assessment processes may be better than older decide-announce-defend systems of hazardous facility siting, more work can be done to better address wider-scale environmental justice.

Stimulating technological fixes

A combination of government regulation and market mechanisms may help encourage environmental technologies. While improved facility siting helps, so too does stimulating the development of new and existing technologies that help abate pollution – either directly (e.g., facility stack scrubbers) or indirectly as a substitute for a more polluting/threatening technology (e.g., solar energy technologies; see Chapter 6). The capitalist system on which the globalized economy depends thrives on constant change and evolution to produce and sell new innovations. Economists have been looking at what economic factors and government policy instruments stimulate investment in pollution reductions (e.g., Fischer et al., 2003). For example, Andreoni and Levinson (2001) find that in the same way there are U-shaped economic (Kuznets) cycles, so too there are U-shaped pollution cycles. Pollution increases and decreases over time for nation states as well as individual industries and products. One of the main predictors of these cycles is the income of the country in which a product is manufactured and sold and the relationship is that pollution decreases as local household incomes rise. This suggests that the more successful a product and the more disposable income available to locals, the greater the investment in pollution abatement for that product. They caution though about inferring that less regulation to let the market improve incomes is the most effective mechanism for stimulating this income to pollution abatement linkage:

> we show that, absent environmental regulations, the pollution-income path may well have an inverse-U shape, but the amount of pollution at every income will be inefficiently high. While it may be reasonable to deduce that at sufficiently high incomes the optimal pollution will be zero, the model in this paper places no limit on the level of income necessary to generate that return. Neither this paper, nor any of its empirical or theoretical predecessors, supports claims that environmental regulations are unnecessary.
>
> (Andreoni and Levinson, 2001, p. 20)

There is much hope that renewable energy, for example, will create the win-win of reducing greenhouse gas (and other pollution) emissions while stimulating the economy. A complex economic input-output model of the European Union suggests that the positive impacts are already being realized. Further, they point out that benefits of deep renewables policies in the EU entails a third win since renewables also reduce domestic reliance on imported fossil fuels to EU states – so-called energy sovereignty (Ragwitz et al., 2009). Such optimistic conclusions are not universal though. Frondel et al. (2010) in their analysis of the economic impact of German renewables policy – a country often touted as being a model for transitioning to green energy – conclude that their current feed-in-tariff strategy is not an efficient path for achieving the three goals of economic growth, energy sovereignty and greenhouse gas reductions. That is, though all are being realized to some degree, there are opportunity costs to the current policy mechanisms. They suggest that all three goals may be achieved more efficiently through other policy levers like emissions trading. This signals a shift in the debate about renewables from "Should we transition to renewables?" to "*How* should we transition to renewables?" Yet the 'how' has become more problematic than perhaps anticipated for such low-pollution technologies. Local social barriers to renewables development such as wind turbines remain an ongoing issue, one which requires less studying from a distance and more proximate, on-the-ground fieldwork in communities living with existing or proposed renewables development. Concerns range from aesthetic changes to the direct health impacts of noise but also the psychosocial well-being impacts of the facility siting process itself – treating communities fairly and paying more attention to the distribution of financial benefits at the local scale (see Figure 10.1; Devine-Wright, 2005).

Paths in all directions?

In the short term it seems that the path followed is as uncertain as the science on which we must base our risk policy decisions in the first place. Scientific uncertainty about the impacts of new and existing chemicals and new technologies will likely remain a problem for some time because scientific assessment of individual chemicals/substances/technologies simply cannot keep pace with current economic growth. The number and length of studies required to get a firm handle on the health threats of any one substance, let alone all suspicious substances, is potentially all consuming (Moeller, 2011). There is still much to potentially be learned though about how laypeople, scientists and policy makers think about and manage risk under uncertainty. For example, although the cultural theory of risk is important for predicting how groups interact on risk, there is much more to be understood about the values that motivate precautionary policy recommendations versus ones that seem to be on a path to repeat past mistakes. The added problem is that context and place matter – for example, DDT and malaria – making it challenging to ascertain what counts as a 'bad' decision path. In this sense a patchwork of different policies for the same substance globally may be entirely rational and reasonable; what is good policy in one context may not make sense in another. Further, what we value (e.g., greenhouse gas reductions, unfettered consumption) may vary over space and time. The (dis)value of loss of life from activities of the nuclear industry is a case in point. Though thousands of deaths may ultimately be attributed to the Chernobyl and Fukushima disasters, analyses of deaths suggest fossil fuel and hydro energy production methods have resulted in as many or more deaths – without fossil fuels having the added benefit of slowing greenhouse gas emissions (see Chapter 9). Do we disvalue radiation-related deaths far more than industrial accidents in mines, on derricks, constructing pipelines and dams as suggested by Kahneman and Tversky's S-shaped utility curve (Chapter 4)? Why? These are examples of entry points for quantitative and qualitative social scientists to further explore risk in ways that are complementary to research based on toxicology and epidemiology.

Is the environment less 'safe'?

Is the world increasingly unsafe due to rising environmental health threats? The book is intentionally not framed around this question, though it is certainly a question that arises within environmental and health studies, particularly in relation to risk. Is our use of technology making things worse or better for our overall health and the environment? At the same time such questions are problematic, fraught with the inevitable definitional dilemmas and debates – for example, what counts as 'safe' (chronic or acute disaster), 'environment' (physical, social, cultural) and even 'health' (disease, mental well-being, community social harmony). There is no doubt that life expectancies are generally higher in industrialized nations compared to their less industrialized counterparts, with the irony being that industrialization and urbanization often lead to short-term health challenges for infectious disease vis-à-vis sanitation and crowding (Cassel and Tyroler, 1961). Yet as indicated in Chapter 2, while largely urban issues of road traffic fatalities have declined in many countries, in many rapidly urbanizing countries road traffic fatalities have gotten worse over recent years. Deciding how to define these things is all value-laden judgement, which is consistent with the idea that environment and health problems and their resolution are value laden from tip to tip – a core theme of this book. This is not meant to suggest we stop studying risk in the lab or using large datasets using the scientific method; instead we are suggesting the values themselves can be explored further.

For others, like Beck and Giddens and the risk society, the questions of whether the world is less safe may be beside the point, since for them it is the perception that it is getting worse and how that socially motivates action is what matters. People will think and potentially act on what they believe to be true, and those 'truths' are at some level tied to deeply held worldviews on the relationship between humans and nature as resilient or fragile and whether the pace of growth and life in industrialized nations is too fast or slow (Brasier et al., 2011).

Hazards appear to be changing, but so too the social values may be changing to adjust to a globalized and cosmopolitan world. As pointed out in Chapters 2 and 4, the dose-response curve extrapolations, cut-points for statistical probabilities and safety levels in toxicology when exposed animals are used to make inferences about humans are expertly defined, but they are still 'defined', which implies value judgements. Neither the science we have tacitly sanctioned in the past nor the hazards we currently face may have fully anticipated new large-scale hazards within globalized systems that carry with them the spectre of uncertain and catastrophic losses.

Policy = (good) science + values

This book concerns environment and health issues from a social science perspective. The theories and frameworks such as the social amplification of risk, risk society and cultural theory of risk discussed at various points in the text have a few common threads. First, we are moving towards a better understanding of how social phenomena relate to risk – for example, values, worldviews, ways of life. Values are infused throughout risk decision making, in particular when individuals and communities must make sense of the science that is available on the environment and health threats associated with a technology, substance or facility. We do not all value the same things equally – for example, protecting local employment opportunities or other benefits may strongly influence how locals think about a proposed new facility or technology. For example, a waste-to-energy facility, in addition to providing jobs, may reduce net greenhouse gas emissions. Groups that emphasize those aspects of the facility may conflict with those more concerned more about fair facility siting and health-harming releases (e.g., heavy metals and dioxins). As highlighted throughout the book, there is typically science to support both sets of concerns. However, not all of that science is necessarily 'good' in the sense that some of it may be based on weak epidemiologic designs; in the risk society such distinctions take time to gain any salience in the risk society if they ever do. A key challenge then for both decision makers and social scientists is sorting out which aspects of the substance/technology/facility are valued – both negatively and positively – and by whom.

A second feature of these frameworks/theories is the idea that risk is, to a *certain extent*, socially constructed. Taken together with the point about values, the socially constructed aspects of risk relate in particular to what people feel should be *done* (or not) about a particular environmental hazard – that is desired courses of action. This combines with the idea that the social processes may strengthen or weaken commitment to particular values (social amplification or attenuation). Strong commitment to the notion that a facility or substance causes irrevocable health threats will then rarely yield to the suggestion that local jobs are a sufficient benefit. Yet the commitment to such a position itself involves a complex set of not-particularly-well-understood relationships between groups. For example, public meetings may help solidify or weaken positions as locals learn where their friends and neighbours stand on the issue. In the absence of face-to-face encounters people may have to rely more on Internet venues – websites – to explore their stance. Eyles et al. (2014) trace various 'conspiracy theories' about events surrounding a flood and fire near a small nuclear power plant in Ft. Calhoun, Nebraska, in 2011. The timing is important because the rising waters of the Missouri River which posed a threat to the plant happened only a few months after the Fukushima disaster. Official attempts to restore calm amid what they felt was 'white noise' of misinformation were met with scepticism via various Internet sources, supporting an overall atmosphere of distrust. Local disputes can have wider consequences though. That is, when such small stories are repeated for other nuclear facilities they connect to larger narratives of distrust in existing institutional arrangements and the official stories of relative safety in the nuclear industry. As social scientists we have only begun to grasp the social processes involved as they relate to Internet media in particular.

Orienting theories like the social constructionism (risk is defined through social interaction) and social evaluation theory (risk is defined in comparison to reference groups) are firmly rooted in the social sciences on risk (Chapter 2). We are not suggesting though that risk is simply relativistic and any and all social constructions must be accounted for in policy. Instead we argue that these theories must be put in the context of due attention to principles like environmental justice and vulnerability as moral, policy and perhaps even scientific guides. Thus, we are not suggesting we throw the proverbial risk assessment baby out with the value- and uncertainty-laden bathwater; instead we argue that we must acknowledge, understand and perhaps embrace these aspects of environmental health. In this sense conventional risk scientists and allied social sciences interested in risk face the same challenges. Who is studied and how they are studied has policy impacts, and this should be attended by constant reflection on the idea that such considerations are necessary.

We have come a long way globally in terms of implementing new tools to manage environment and health threats including updated toxicological testing, widespread environmental legislation including guidelines for environmental assessment and voluntary facility siting. These are thanks in part to global cooperation on environmental issues in such guises as the Rio Summit and the Montreal and Kyoto Protocols. Yet as discussed in Chapter 5 some of these, like the Kyoto Protocol, have not met many of the rather ambitious targets – in the Kyoto case, to minimize the impacts of global climate change. Greenhouse gases and the other pollutants that are associated with fossil fuels and other forms of consumption and waste are still rising despite the great strides that have been made. While places like the European Union are legislating to transition to renewable energy, other significant economies including China, the U.S. and Canada have considerable challenges ahead for reducing the use of coal and oil for energy.

With local resistance to renewables there will be ongoing challenges to satisfy environmental targets and simultaneously respect the advances we have made in thinking about environmental justice. For example, though a decide-announce-defend model for facility siting works for placing technologies like wind turbines and hydro dams, they violate core principles for fair/just environmental planning. The greater good need not come entirely at the expense of locals, who may be cautiously willing to accept jobs and other facility benefits to sustain their communities. The formidable task is finding ways to negotiate those arrangements so they are not perceived as 'bribery' or 'blood money' by locals or 'gold-digging' by developers, governments and non-local publics. Community impact-benefit agreements including community profit sharing are related ways to simultaneously address the greater environmental health good and properly acknowledge the negative externality risks taken on by locals. The resource industry has also taken on the allied notion of *social licence*, a process that acknowledges the scalar politics of developments – that the local community must bear a disproportionate share of negative impacts:

> Social licence is premised on the idea of informal or 'tacit' licensing that signals the presence or absence of a critical mass of public consent, which may range from reluctant acceptance to a relationship based on high levels of trust. By definition, this licence is considered to be fundamentally intangible and informal, unless effort is made to measure, analyse or quantify its character.
>
> (Owen and Kemp, 2013, p. 31)

As indicated throughout this volume, however, the trust that is required for such agreements between industry and local communities is difficult to establish and likely involves moving the dialogue away from technically oriented risk analysis to some degree. Thus, one way forward is to create spaces for public dialogue on a wide range of environmental health issues that includes ways of life and well-being. Again, social science can play a significant role in making sense of these processes as they unfold and monitor their impacts using appropriate methodologies.

Compelling evidence, dubious action: Electromagnetic fields

There are a number of environmental exposures that are not necessarily tied to a specific facility though, and policy for these can be equally problematic. This is sometimes where values about how we live potentially come into tension with desires for high safety. The case of electromagnetic fields (EMFs) and health has become a curious mix of both with the advent of the cell phone and other handheld devices which emit radio-frequency electromagnetic fields (RF-EMFs). While the concern about EMFs tended to centre around childhood leukemia from overhead wires and power stations – static infrastructure – RF-EMFs are generally from consumer products. The epidemiologic work on this topic is very extensive and research grew up in the 1970s and 1980s with a mixture of findings, some showing statistically significant impacts and others showing no effect. Authors like Savitz et al. (1989) have pointed out the extreme difficulty of determining causal association because of the problem of confounding – EMFs are essentially 'everywhere' in the industrialized world. Further, they pointed out that there was no plausible biological link between EMFs and cancer (see also Knave, 2001). On the flip side of the scientific uncertainty coin for this exposure, it is because of the ubiquity of exposure that epidemiologists are capable of designing studies with the statistical power robust enough to detect rare outcomes – including cancers like leukemia. That is, within these large sample studies there are lower fears of false negative results, as sample sizes are sufficiently large to be powerful enough to detect effects if they are indeed present – that is, true positives (see Chapter 1). In fact many of the studies are case-control; thousands of those with leukemia are matched and compared for exposure differences to similar people who do not have the disease (e.g., Draper et al., 2005).

With this enhanced capacity for detecting effects, a number of researchers have concluded that indeed there is a statistically significant effect of EMFs and RF-EMFs, particularly on certain forms of cancer. For example, Draper et al. find that within a group of 9,700 children who had been diagnosed with leukemia, they were more likely to live within 200 m of overhead high voltage power lines compared with non-leukemia children living greater than 600 m away. The relative risk is reasonably small at 1.7, and they conclude that 1% of leukemia cases in England and Wales might then be attributable to high-voltage lines (since most do not live within 200 m). Others have conducted systematic reviews or pooled studies of collections of studies to conclude that the impact of EMFs is significant but only accounts for a relatively small proportion of leukemias. For example, Greenland et al. (2000) study the findings from 15 studies and conclude that the relative risk between higher-exposed and lower-exposed children is also 1.7 but that in the U.S. context this implies that EMFs may cause as high as 8% of childhood leukemias.

As far as RF-EMFs are concerned a panel of experts working on behalf of the International Agency for Research on Cancer (IARC) concluded that RF-EMF exposure from devices such as cell phones represent a "possible cause of cancer" – the 2B IARC classification – which is in the middle of their fivefold[2] classification system which ranges from 1 "carcinogenic to humans" (e.g., alcohol, radon, asbestos, benzene) to 4a "probably not carcinogenic to humans" (Baan et al., 2011). The panel reviewed all of the toxicological and epidemiologic research to date on RF-EMFs to come to this conclusion.

This seems to be a case of dooming ourselves by exposing ourselves and our children to a possible carcinogen – whither the precautionary principle? The EMF and RF-EMF research combined is about as compelling as it gets in the world of risk assessment, yet the causes of these cancers are becoming more rather than less ubiquitous. Yet the same can be said of a range of other substances including alcohol. As we have indicated in Chapter 4 the voluntariness of the exposure can have a profound impact on how the risk is perceived, with imposed infrastructure more prone to being feared while the risks of cell phones get downplayed by individuals since they have the capacity to choose their own exposure. Likewise, the perceived benefits of electricity and cell phones are tacitly understood to warrant the risk. This suggests a societal conundrum in the sense that the value and promise of exciting new technologies may trump precautionary approaches. This might readily answer the question, "If we knew then what we know now about EMFs and RF-EMFs and cancers, would we still allow them to be developed and sold?" The answer is likely a qualified yes with efforts nevertheless taken to minimize risk by, for example, building homes away from power lines and warning cell phone users to use hands-free microphones. Yet as the risk society theory suggests, we are less prone to tacitly accept risks from 'new' technologies like nuclear energy and genetically modified organisms.

Conclusion

In the process for deciding on a by-law for banning cosmetic pesticides in Halifax, Nova Scotia, Canada, the City Council's committee tasked with drafting the details made the bold decision to set the scientific data to one side and instead focus on how a by-law could practically be implemented. On the surface, it may seem like the decisions they eventually made ignored the science. Instead, what they were acknowledging was that the science was mixed and ultimately equivocal – not only would they be considering a whole range of cosmetic pesticides, they would also be considering a wide range of potential health outcomes. These considerations had the potential to stymie the committee and ultimately support the status quo of ongoing household chemical pesticide use. Thus, their policy was overtly precautionary and overtly value based – risks to children in particular were deemed too great amid considerable scientific uncertainty (Hirsch et al., 2010).

Are we doomed to make the same mistakes if we place all of our faith in positivistic approaches to risk assessment and management to ostensibly let the data speak for themselves? Our response is a qualified yes, but this answer is meant to open up spaces for understanding rather than close them. The data more often do not speak for themselves and further, analysis is attended by a series of decisions that are in many ways value laden. How we interpret the data from toxicological and epidemiological studies is even more explicitly inscribed with values than we have admitted in the past. This does not mean society should retreat to an anti-science stance; foundational risk sciences absolutely have a place in ongoing assessments of substances, technologies and facilities. Instead we have cast light in the corners of risk assessment and management for environment and health – to spotlight places where social scientists can play a role in helping us acknowledge and sort through the values that support various views and actions on risk. While differences between 'expert' and 'lay' views on risk are interesting, it might be more useful for sound policy and action to think in terms of coalitions in support of and concerned about the risks from technologies and facilities. If people are being asked to take on risks in the name of the public good, sound policy suggests that it is indeed an 'ask' and not a 'tell'; and accepting risks in willingly and in an informed manner is overtly social and usually politicized. Those concerned that certain hazards pose unacceptable risks are not necessarily uninformed. In fact, the opposite is often the case, whereby these people and the coalitions with which they are aligned inform themselves of the details of all manner of scientific enquiry and in some cases participate in producing their own evidence. Yet that evidence is likely to have enough gaps to allow sufficient support for a number of risk policy positions. Framing these positions as cultures of risk that are at some level socially constructed should be useful for both policy change and learning as well as social scientific research.

Notes

1 After a decades long siting process, the town of Forsmark has been selected, but the beginning of construction is not planned until sometime in the 2020s (SKB, 2015).
2 Though the numbering is from 1 to 4, classification 2 has two subsections: 2a (probably carcinogenic to humans), and 2b (possibly carcinogenic to humans).

References

Ackerman, F., & Heinzerling, L. 2002. Pricing the priceless: Cost-benefit analysis of environmental protection. *University of Pennsylvania Law Review*, 150(5), 1553–1584.

Alhakami, A. S., & Slovic, P. 1994. A psychological study of the inverse relationship between perceived risk and perceived benefit. *Risk Analysis*, 14(6), 1085–1096.

Andreoni, J., & Levinson, A. 2001. The simple analytics of the environmental Kuznets curve. *Journal of Public Economics*, 80(2), 269–286.

Baan, R., Grosse, Y., Lauby-Secretan, B., El Ghissassi, F., Bouvard, V., Benbrahim-Tallaa, L., . . . & Straif, K. 2011. Carcinogenicity of radiofrequency electromagnetic fields. *The Lancet Oncology*, 12(7), 624–626.

Bansal, R. 2009. AP-S turnstile: Say au revoir to cellphones. *IEEE Antennas Propagation Magazine*, 51, 152.

Baxter, J., & Greenlaw, K. 2005. Revisiting cultural theory of risk: Explaining perceptions of technological environmental hazards using comparative analysis. *Canadian Geographer*, 49(1), 61–80.

Baxter, J., & Lee, D. 2004. Explaining the maintenance of low concern near a hazardous waste treatment facility. *Journal of Risk Research*, 6(1), 705–729.

Bissett, K. 2015. *Canadian Nuclear Power Plants Completing Upgrades Prompted by Fukushima Disaster.* Available at: http://globalnews.ca/news/2118533/canadian-nuclear-power-plants-completing-upgrades-prompted-by-fukushima-disaster/ (Accessed 27 July 2015).

Brasier, K.J., Filteau, M.R., McLaughlin, D.K., Jacquet, J., Stedman, R.C., Kelsey, T.W., & Goetz, S.J. 2011. Residents' perceptions of community and environmental impacts from development of natural gas in the Marcellus Shale: A comparison of Pennsylvania and New York cases. *Journal of Rural Social Sciences*, 26(1), 32–61.

Broughton, E. 2005. The Bhopal disaster and its aftermath: A review. *Environmental Health: A Global Access Science Source*, 4(6), 1–6.

Cassel, J., & Tyroler, H.A. 1961. Epidemiological studies of culture change: I. Health status and recency of industrialization. *Archives of Environmental Health: An International Journal*, 3(1), 25–33.

CBC News. 2015a. *Phosphine Poisoning: 2nd Child Dies in Edmonton Hospital.* Available at: http://www.cbc.ca/news/canada/edmonton/phosphin e-poisoning-2nd-child-dies-in-edmonton-hospital-1.2974258 (Accessed 2 March 2015).

CBC News. 2015b. *Phosphine Gas Likely Cause of Thailand Deaths of Quebec Sisters: Coroner.* Available at: http://www.cbc.ca/news/canada/montreal/phosphine-gas-likely-cause-of-thailand-deaths-of-quebec-sisters-coroner-1.2977667 (Accessed 2 March 2015).

Centre for Research on the Epidemiology of Disasters (CRED). 2015. *The International Disaster Database.* Available at: http://www.emdat.be/ (Accessed 25 February 2015).

Chapman, S., George, A.S., Waller, K., & Cakic, V. 2013. The pattern of complaints about Australian wind farms does not match the establishment and distribution of turbines: Support for the psychogenic, 'communicated disease' hypothesis. *PloS One*, 8(10), e76584.

Crichton, F., Chapman, S., Cundy, T., & Petrie, K.J. 2014. The link between health complaints and wind turbines: Support for the nocebo expectations hypothesis. *Frontiers in Public Health*, 2, 220–236.

Devine-Wright, P. 2005. Beyond NIMBYism: Towards an integrated framework for understanding public perceptions of wind energy. *Wind Energy*, 8(2), 125–139.

Dhara, V.R., & Dhara, R. 2002. The union carbide disaster in Bhopal: A review of health effects. *Archives of Environmental Health: An International Journal*, 57(5), 391–404.

Draper, G., Vincent, T., Kroll, M.E., & Swanson, J. 2005. Childhood cancer in relation to distance from high voltage power lines in England and Wales: A case-control study. *British Medical Journal*, 330(7503), 1290.

Eyles, J., Fried, J., & Eyles, E.C. 2014. Redacted in Nebraska: The noises of conspiracy around nuclear power problems. *GeoJournal*, 79(3), 329–342.

Festinger, L. 1957. *A Theory of Cognitive Dissonance.* California: Stanford University Press.

Fintel, B., Samaras, A., & Carias, E. 2009, July 28. The thalidomide tragedy: Lessons for drug safety and regulation. *Helix.* Available at: https://helix.northwestern.edu/article/thalidomide-tragedy-lessons-drug-safety-and-regulation (Accessed 25 February 2015).

Fischer, C., Parry, I.W., & Pizer, W.A. 2003. Instrument choice for environmental protection when technological innovation is endogenous. *Journal of Environmental Economics and Management*, 45(3), 523–545.

Fontevecchia, A. 2013, Feb 5. BP fighting a two front war as Macondo continues to bite and production drops. *Forbes*. Available at: http://www.forbes.com/sites/afontevecchia/2013/02/05/bp-fighting-a-two-front-war-as-macondo-continues-to-bite-and-production-drops/ (Accessed 25 February 2015).

Foster, K.R., Vecchia, P., & Repacholi, M.H. 2000. Science and the precautionary principle. *Science*, 288(5468), 979–981.

Frondel, M., Ritter, N., Schmidt, C.M., & Vance, C. 2010. Economic impacts from the promotion of renewable energy technologies: The German experience. *Energy Policy*, 38(8), 4048–4056.

Greenland, S., Sheppard, A.R., Kaune, W.T., Poole, C., Kelsh, M.A., & Childhood Leukemia-EMF Study Group. 2000. A pooled analysis of magnetic fields, wire codes, and childhood leukemia. *Epidemiology*, 11(6), 624–634.

Havenaar, J., Rumyantzeva, G., Kasyanenko, A., Kaasjager, K., Westermann, A., van den Brink, W., . . . & Savelkoul, J. 1997. Health effects of the Chernobyl disaster: Illness or illness behavior? A comparative general health survey in two former Soviet regions. *Environmental Health Perspectives*, 105(Suppl 6), 1533.

Hirsch, R., & Baxter, J. 2009. The look of the lawn: Pesticide policy preference and health-risk perception in context. *Environment and Planning C: Government & Policy*, 27(3), 468.

Hirsch, R., Baxter, J., & Brown, C. 2010. The importance of skillful community leaders: Understanding municipal pesticide policy change in Calgary and Halifax. *Journal of Environmental Planning and Management*, 53(6), 743–757.

Keystone XL and the president's veto: Fuelling anger. (2015). *The Economist*. Available at: http://www.economist.com/blogs/democracyinamerica/2015/02/keystone-xl-and-presidents-veto (Accessed 25 February 2015).

Knave, B. 2001. Electromagnetic fields and health outcomes. *Annals of the Academy of Medicine, Singapore*, 30(5), 489–493.

Lyons, R.A., Temple, J.M., Evans, D., Fone, D.L., & Palmer, S.R. 1999. Acute health effects of the Sea Empress oil spill. *Journal of Epidemiology and Community Health*, 53(5), 306–310.

Matlack, C. 2013, April 18. Texas fertilizer explosion recalls French disaster, *Bloomberg Business*. Available at: http://www.bloomberg.com/bw/articles/2013–04–18/texas-fertilizer-explosion-recalls-french-disaster (Accessed 25 February 2015).

McLaughlin, E. 2013, April 22. West, Texas, fertilizer plant blast that killed 15 'preventable,' safety board says. *CNN*. Available at: http://www.cnn.com/2014/04/22/us/west-texas-fertilizer-plant-explosion-investigation/ (Accessed on 25 February 2015).

Moeller, D.W. 2011. *Environmental Health*. Boston: Harvard University Press.

Obama oil spill commission's final report blames disaster on cost-cutting by BP and partners. 2011, Jan. 5. *The Telegraph*. Available at: http://www.telegraph.co.uk/finance/newsbysector/energy/oilandgas/8242557/Obama-oil-spill-commissions-final-report-blames-disaster-on-cost-cutting-by-BP-and-partners.html (Accessed 25 February 2015).

Owen, J.R., & Kemp, D. 2013. Social licence and mining: A critical perspective. *Resources Policy*, 38(1), 29–35.

Page, L., Petrie, K., & Wessely, S. 2006. Psychosocial responses to environmental incidents: A review and proposed typology. *Journal Psychosomatic Research*, 60, 413–422.

Ragwitz, M., Schade, W., Breitschopf, B., Walz, R., Helfrich, N., Rathmann, M., et al. 2009. *EmployRES. The Impact of Renewable Energy Policy on Economic Growth and*

Employment in the European Union. Available at: http://ec.europa.eu/energy/renewables/studies/doc/renewables/2009_employ_res_report.pdf (Accessed 11 May 2016).

Rehman, W., Arfons, L.M., & Lazarus, H.M. 2011. The rise, fall and subsequent triumph of thalidomide: Lessons learned in drug development. *Therapeutic Advances in Hematology*, 2(5), 291–308.

Savitz, D.A., Pearce, N.E., & Poole, C. 1989. Methodological issues in the epidemiology of electromagnetic fields and cancer. *Epidemiologic Reviews*, 11(1), 59–78.

Shepherd, D., McBride, D., Welch, D., Dirks, K.N., & Hill, E.M. 2011. Evaluating the impact of wind turbine noise on health-related quality of life. *Noise and Health*, 13(54), 333.

Sjöberg, L. 2004. Local acceptance of a high-level nuclear waste repository. *Risk Analysis*, 24(3), 737–749.

SKB. 2015. *How Forsmark Was Selected.* Available at: http://www.skb.com/future-projects/the-spent-fuel-repository/how-forsmark-was-selected/ (Accessed 27 July 2015).

Sundqvist, G. 2002. *The Bedrock of Opinion: Science, Technology and Society in the Siting of High-Level Nuclear Waste* (Vol. 32). Berlin: Springer Science & Business Media.

Todt, O., & Lujan, J.L. 2014. Analyzing precautionary regulation: Do precaution, science and innovation go together? *Risk Analysis*, 34(2), 2163–2173.

Vaughan, D. 2009. Slopes, repeating negative patterns, and learning from mistake? In Starbuck, W., & Farjoun, M. (eds.). *Organization at the Limit: Lessons from the Columbia Disaster*. Maldern, MA: BLackwell, pp 41–59.

Walker, C., Baxter, J., & Ouellette, D. 2014. Adding insult to injury: Psychosocial stress in Ontario wind turbine communities. *Social Science & Medicine*, 2(4), 424–442.

WHO. 2005. Health Effects of the Chernobyl Accident and Special Health Care Programmes: Report of the UN Chernobyl Forum Expert Group 'Health' (EGH). Working draft. World Health Organization, Geneva (Switzerland).

Index

Page numbers in *italics* refer to figures and tables.